Responsible Tourism: Concepts, Theory and Practice

Responsible Tourism: Concepts, Theory and Practice

Editor

Oliva Cabral

scitus
academics

Responsible Tourism: Concepts, Theory and Practice
Edited by **Oliva Cabral**

ISBN: 978-1-68117-621-5
Library of Congress Control Number: 2016946479

616, Corporate Way, Suite 2, 4766,
Valley Cottage, NY 10989

www.scitusacademics.com

Notice

Printed in United States of America on Acid Free Paper

Preface

Tourism is one of the most important industries in the world. Responsible tourism is concerned about the impact of tourism on communities, ecology and people and aims to reduce these impacts by providing tourism that benefits host communities, improves working conditions, involves the community in the region, promotes cultural heritage and benefits the environment. To put it simply, responsible tourism is tourism, which creates far better places for people to live and visit. It generates greater financial benefits for local people and improves the health of local communities. It also contributes positively to the preservation of cultural and natural heritage and to the preservation of the diversity of the world.

This book contains leading topics on responsible tourism all together, including tourism policy planning and management, wildlife and heritage tourism. This book is suitable for university libraries, policy-makers and sustainable tourism researchers. Responsible tourism is a tourism management approach designed to maximize economic, environmental and social benefits and reduce costs for destinations. A responsible approach to tourism aims at achieving the triple end results of sustainable development. The distinctive feature of the strategy is the obligation of stakeholders in the tourism sector and destinations in general to take action to achieve sustainable tourism development.

Table of Contents

1

CHAPTER

Sustainable-Responsible Tourism Discourse – Towards 'Responsustable' Tourism

Tanja Mihalic

University of Ljubljana, Faculty of Economics, Slovenia

HIGHLIGHTS

- Defines the dominant tourism discourse on *sustainability* and *responsibility*.
- Defines the *sustainability* as a theory (a concept); defines the *responsibility* as practice (appropriate action).
- Develops a Triple-A Model.
- Suggests and justifies the new tourism nomination: *responsustable tourism*.
- Complements the sustainable indicators with Triple-A Model.

ABSTRACT

Despite several decades of academic and practical debate on tourism sustainability, its application in practice remains difficult. The dominant tourism discourse on *sustainability*(theory, seen as a concept) and *responsibility* (practice, understood as appropriate action) calls for a solid understanding of the process of how a responsible destination actually implements a sustainability agenda, which this paper aims to provide. In this context, we explore theoretical perspectives from political economics and behavioural economics to offer a well-reasoned integrated sustainability–responsibility model comprising three stages: Awareness, Agenda and Action. This Triple-A Model complements the sustainability indicators debate and provides advice on how to continuously implement the sustainability concept and move from market-value-led and environmentally laissez-faire tourism towards more environmental- and social-value-driven responsible tourism. In addition, this paper discusses the existing sustainability and responsibility nomenclatures and their use and contributes relevant conclusions on the current understanding of sustainability and responsibility in European and UNWTO practices. The term *responsustable tourism* is suggested to join two existing terms and demonstrate that the current understanding of responsible tourism behaviour is based on the concept of sustainable tourism.

Keywords: Sustainable tourism; Responsible tourism; Responsustable tourism; CSR; Behavioural economics; Triple-A Model

1. INTRODUCTION

Concern over the natural and social environments has generated research debate on tourism–environment relationship. This debate gained momentum in the early 1970s when George Young argued that the impacts of tourism are both a blessing and a blight (Young, 1973) and Claude Kaspar, a Swiss-based tourism researcher, called for a new "dimension of [the] tourism debate" (Kaspar, 1973, p. 139), which he termed*environmental ecology*. Later, Swiss ecologist Jost Krippendorf (1984) challenged the sense of mass tourism in his book entitled "Vacation People" ("Ferienmenshen" in German, translated as "Holiday Makers" in English (Krippendorf, 1987)) and began the search for alternatives. Tourism critics in the 1980s called for "... more responsibility for the effects of travel and behaviour on host environments, both physical and human" (Butler, 1995, p. 5). This interest in "more responsibility" led to so-called alternative tourism forms and concepts which have been given many names, such as alternative, soft, quality, eco, responsible, minimum impact tourism, green and ethical tourism, with all of them representing an alternative to the mainstream mass tourism that has been becoming environmentally, socially, ethically and politically intolerable (Mihalič,

2006 and Swarbrooke, 1999). In general, global environmental concern culminated in "Our Common Future", as defined in the Brundtland report (WCED, 1987), which brought the global environmental debate and the notion of sustainability to the forefront of global and local social and political thinking and agendas. The tourism sustainability debate, e.g., the debate regarding use of the term *sustainability*, following the "Our Common Future" legacy began in the early 1990s with Edward Inskeep (1991) who defined five main criteria for *sustainable tourism*, which addressed the economic, environmental and social responsibility of tourism as well as its responsibility towards tourists (visitor satisfaction) and global justice and equity. Some of his criteria received little recognition in the following debate, which also originated in the Brundtland legacy and culminated in the next decade. The United Nations' organisations, including the UNWTO, primarily supported the three-pillar (environmental, socio-cultural and economic) concept of sustainable tourism. This concept became the focus of mainstream academic tourism literature and programmes and the input of many tourism strategies and policies, resulting in a recognised global trend towards sustainable tourism development. However, sustainable tourism research, documents and actions have recently been increasingly accompanied by the notion of responsible tourism. Examples include the new European document entitled the "Charter for Sustainable and Responsible Tourism" (TSG, 2012), and two recent books called "Responsible Tourism" (Leslie, 2012a) and "Taking Responsibility for Tourism" (Goodwin, 2011).

The trend towards sustainability has been studied and accepted by many researchers. On one hand, the sustainability concept has served for some as a magic wand pointing towards more sustainable, environmentally and socially friendlier tourism developmental models and forms (Ritchie and Crouch, 2000 and Swarbrooke, 1999). On the other hand, the concept has been persistently criticised for being flawed and inadequate (Higgins-Desbiolles, 2010). In other words, the sustainability discussion has helped draw attention to the need for a balance between economic and environmental interests in tourism. Its actual penetration into strategies and policies has resulted in many good practices and improvements such as energy savings, recycling, a reduction of waste and emissions and attempts to improve the livelihood of the local population. However, there is also significant evidence of the opposite effect. Wheeler argues that the "intellectually appealing" concept of sustainable tourism has little practical application because it has turned into a public relations tool for addressing the criticism of the impact of tourism while allowing essentially the same behaviour as before (Wheeller, 1993, p.121). Indeed, a consensus on the efficacy of sustainable tourism development remains elusive (Chettiparamb and Kokkranikal, 2012), and its implementation in practice remains difficult, leaving much of the tourism industry "... alarmingly unsustainable" (Higgins-Desbiolles, 2010, p. 117).

Tourism stakeholders are applying sustainability practices at a slow pace. Further, some destinations might inaccurately promote themselves as sustainable and increase the expectations of new tourists (Poon, 1989), who are then confronted with the gap between the actual and promised (advertised) sustainability. Extensive lists of sustainable tourism indicators have been created to measure this gap and actual sustainability (EC., 2014,ECETAT and ECOTRANS, 2004, TSG., 2007 and UNWTO., 2004). However, although these lists enable actual progress in sustainability performance to be monitored while some (aside from the three-pillar approach) measure political sustainability and customer satisfaction (EC., 2014 and ECETAT and ECOTRANS, 2004), they do not provide a measurement tool that can help destinations understand the overall transition process regarding sustainability and responsibility. For this reason, a tool to understand, measure and monitor the process of implementing sustainability is still needed. This is the primary purpose of this paper.

The existence of this paper has been provoked by two inter-related facets of the above debate. The first relates to the never-ending search for new tourism terms on the assumption that a new term will bring *more responsible tourism*. In this context, interest has been triggered by the recently growing popularity of the term *responsible tourism* in the tourism literature (Goodwin, 2011) as well as in consulting and political and business practices (TSG, 2012). The actual and academic coexistence of these two key terms can also be well illustrated by the titles of scientific conferences discussing sustainable-responsible tourism development. For example, in early October 2013, on the same days, two conferences were held on the issue, one in Istanbul, Turkey and the other in Barcelona, Spain. The first conference tried to attract participants through the conference title "sustainability" (International conference: Sustainability issues and Challenges in Tourism, 2013), the other by discussing "responsible tourism" (RTD7 Conference: Responsible Tourism in Destinations. Barcelona – Catalunya, 2013). Questions arise as to what does responsible tourism bring to the tourism debate in terms of terminology, concept and tourism type, and how does it relate to sustainable tourism.

The second facet relates to the gap between the appealing conceptual idea of sustainable tourism and its alarmingly slow penetration of action and practice, which is obviously connected to *tourism irresponsibility* or irresponsible tourism behaviour. In the context of both facets, this paper is interested in the penetration of sustainability and responsibility in the tourism industry and destination practices. Being a conceptual paper, its aim is to integrate existing theoretical and practical understandings and uses of the notions of sustainable and responsible tourism and provide new theoretical perspectives for their well-reasoned and coherent understanding. It examines the theories on the causes of environmental damage and the transition of society to a state in which environmental issues can no longer be ignored, as outlined by the Swiss welfare economist Bruno Frey (1985). His theory is applied to tourism to increase the understanding of the sustainable-responsible tourism discourse and current developments in use of the terms sustainable and responsible tourism. The increased use of the term *responsible tourism* is discussed, and an attempt to connect it with*sustainable tourism* is made. Paradoxically, the criticism of the above-mentioned never-ending renaming of tourism has led to a new tourism term. The term *responsustable tourism* has been suggested, not to offer a new tourism type or concept but as an attempt to join two existing terms to properly articulate the current *responsible tourism* debate, which is based on the concept of *sustainable tourism*.

In order to follow the primary purpose of this study, there are three specific aims: first, to understand the notions of responsible and sustainable tourism in a historical, theoretical and practical context; second, to provide a logical model to accommodate the above understandings; and third, to develop a tool to allow actual understanding and implementation of the sustainable tourism concept in a (responsible) destination.

Accordingly, the section "Introduction" explains the relevance and importance of the study. The next section explains the nature of the paper and the methodology applied. The state of the art on the sustainable-responsible tourism discourse is then presented and theoretical and practical evidence of its existence is given. Based on this discourse, the next section provides a model for responsustable tourism. Results and a discussion of the current and a proposed understanding of the sustainable-responsible tourism discourse follows, while the paper finishes with a section outlining conclusions and further research.

2. METHODOLOGY

In line with the standards for conceptual or review articles (Watts, 2011), this paper attempts to further expand and refine the understanding of the sustainable-responsible tourism discourse and suggests how to close the sustainable-responsible gap by using clear definition of each term, derived from theories on environmental damage and behavioural economics.

However, although primarily a conceptual paper, the paper's construction is informed by some explorative methods in theorising and conducting research. In this context, the paper applies engaged scholarship and action research methodology.

Engaged scholarship research emphasises advances in scientific and practical knowledge (Van de Ven, 2007) which fits the theoretical and practical nature of the present research. Being based on ongoing research and academic thinking on sustainable and responsible tourism development, this research engages academic knowledge. Further, by deriving from the current practices in strategy and policy formulation and the implementation of sustainable and/or responsible tourism, it engages with reality. Both are iterated to fit the proposed model.

In addition, the paper employs an action research methodology. Zuber-Skerritt (1996)suggests that the aim of any action research project is to bring about the practical improvement or development of social practice, and practitioners' better understanding of their practices. A key feature of this action research approach is that it involves the investigation being undertaken by the person directly concerned with the social situation under consideration (McGugan, 2002), here derived from the researcher's experience with academic research and the creation and implementation of national and European sustainable tourism strategies and policies. It means that this methodology not only focuses on the transfer of knowledge and dissemination for the purpose of adding to the existing body of tourism knowledge but, most importantly, critical thinking about this knowledge and the co-creation of new knowledge are enhanced by learning from both sides (academic and practical), whether through academic papers, conferences or work experience with the tourism industry and public and private sector in designing sustainable tourism development strategies, agendas and monitoring the implementation process.

The sustainable-responsible tourism discourse is brought to a close by the proposed Triple-A Model which joins sustainable and responsible tourism to form responsustable tourism. Some metaphoric examples of climbing a mountain are used by way of inspiration to interpret, visualise and concretise the theoretical model, its phases and penetration processes. The result of this conceptual thinking is jointly summarised in the section entitled *Framework for understanding the sustainable-responsible discourse*which fully illustrates how to shape that discourse.

3. THE SUSTAINABLE-RESPONSIBLE TOURISM DISCOURSE

In one study, Higgins Desbiolles (2010) claimed that a dominant issue in the current tourism discourse is the elusiveness of achieving sustainability.

Although she argues that the sustainability concept is not adequate, the discourse she refers to actually relates to the low application of sustainability principles in the tourism industry, which "... remains alarmingly unsustainable" (Higgins-Desbiolles, 2010, p. 117). For much of the tourism industry, overall sustainability practices seem as distant as the mighty Mount Everest seems to mountain climbers. Nevertheless, Everest is on almost every professional climber's dream agenda and, although many alpinists have conquered it so far, the majority of mountain climbers have not yet - and never will - reach even its base camp for a wide variety of reasons. In this context, this paper further expands the tourism discourse by distinguishing between dream and reality agendas and actions, e.g., between sustainable and responsible tourism, which is observed as a discourse between the concept and its actual application.

Thus, the first theme of this section is sustainability as a concept. Tourism researchers (Bramwell and Lane, 1993) have traced the theoretical origins of sustainable tourism back to a publication in 1973 (Dasmann et al., 1973) and its political penetration in 1980 in the World Conservation Strategy (IUCN, 1980). As already pointed out in the "Introduction" section of this paper, the drive towards sustainable tourism has been prompted by concerns about the tourism–environment relationship as well as interest in developing concepts that might help bring this relationship more in line with the ideology of "Our Common Future" (WCED, 1987). The debate is focused on the new developmental principles based on the concept termed *sustainability*. It is beyond dispute that sustainable tourism development has been widely recognised as a *concept*, meaning "... tourism that is based on the principles of sustainable development" (UNEP & WTO, 2005, p. 11). These principles, also known as the three pillars, demand consistency with economic as well as socio-cultural and environmental objectives and enable tourism operations to remain competitive in the long run. In this context, some academics refer to economic, socio-cultural and environmental sustainability. However, feeling that it is potentially critical for the acceptability of any solution, some critics tried to integrate a fourth dimension - political sustainability (Ritchie and Crouch, 2000). Yet since it was impossible to define this political dimension as the fourth pillar, because - unlike the economic, socio-cultural and environmental pillars - it does not relate directly to the impacts of tourism, the fourth pillar of the sustainability concept has never materialised. However, parallel to this debate, some addressed the political dimension through a debate on three requirements that must be satisfied to make tourism sustainable (Fig. 1). First, sustainable tourism is based on an awareness of sustainability and ethics, supported by environmental education and information for all stakeholders on both the demand and supply sides. Second, it also relates to the above-mentioned political dimension as it requires the informed participation of all relevant destination stakeholders, a consensus, a critical mass and strong political leadership to enable its implementation. Third, sustainable tourism should maintain a high level of tourist satisfaction, thereby meeting market needs (Mihalič, 2013 and UNWTO., 2004). More specifically, in existing market economies the latter also assumes a willingness to pay for sustainability.

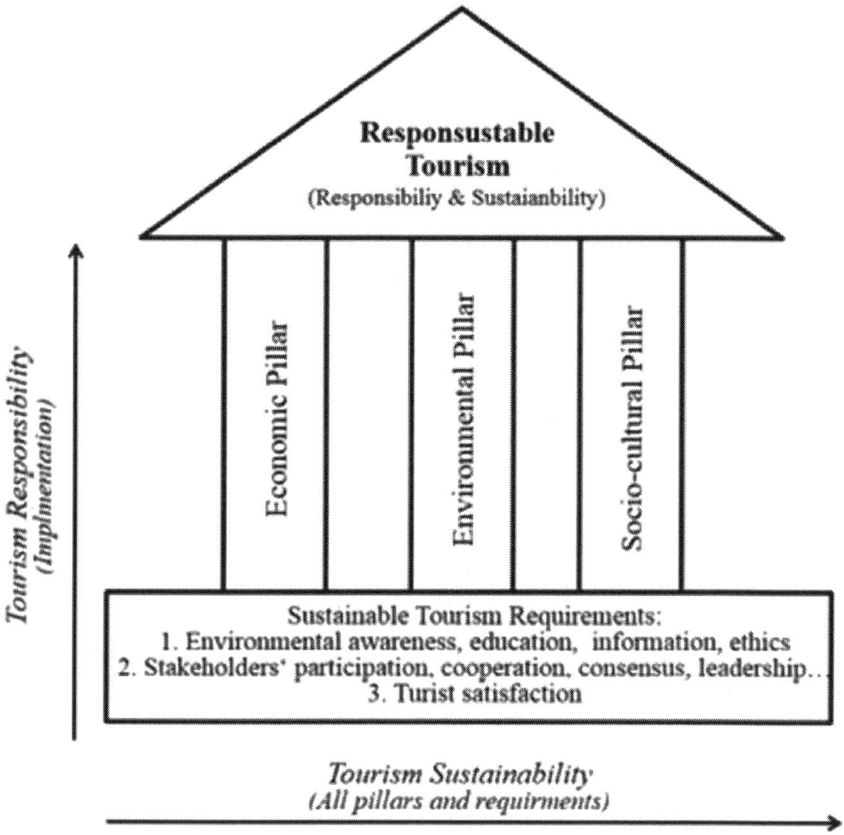

Figure 1. Proposed understanding of responsustable tourism (based on the three pillars and three requirements).

The three-pillar concept has inspired most sustainability research and penetrated numerous strategy and policy documents in the real tourism sector. However, the question remains: to what extent do sustainability principles improve the sustainability of the real tourism sector?

Notably, the tourism sustainability debate has focused on the destination level rather than the micro-level of the firm (Dwyer, 2005). Empirical evidence from the hospitality industry has shown that economic performance is the highest priority, while environmental performance is the lowest priority (Blackstock et al., 2008, Bohdanowicz et al., 2005,Bramwell et al., 2008 and Mihalič et al., 2012). It appears that tolerating and accepting sustainability agendas has been easier for public stakeholders than for private stakeholders, especially corporate businesses since they traditionally focus on economic goals. More specifically, firms have been much slower to discuss or address the environmental and socio-cultural aspects of their business practices and reluctant to talk about *sustainability*. However, other sustainability-coloured concepts such as the *triple bottom line* (TBL) or *corporate social responsibility* (CRS) seem to be closer to the business world. Reluctance to talk about sustainability has also lead to the use of notion*green tourism*, as a synonym.

Therefore, the second theme of this section involves identifying why companies have accepted CSR models more widely. To understand how firms perceive responsibility, two alternative lines of thought known as the customer and broader stakeholder orientations of the firm are relevant (Heikkurinen and Bonnedahl, 2013). According to first line of thought, the firm is market-oriented and "its social responsibility ... is to increase its profits" (Friedman, 1962, citied in van Marrewijk, 2003, p. 97). Profit is the focal point of the company, while social responsibility and actions lie in the domain of the government. In this regard, companies are interested in CSR behaviour only to the extent that it contributes to the business goal, which is ultimately driven by profit motives (Chettiparamb and Kokkranikal, 2012). This market-oriented view can be demonstrated by the argument frequently used by the tourism industry to counter larger environmental pro-activeness. Firms often claim that low consumer demand and low willingness to pay extra for environmentally friendlier products and responsible tourism practices is an insufficient incentive to improve environmental performance (Budeanu, 2005). Market orientation explicitly focuses on the creation of superior value for buyers and thus superior performance for the business and distinguishes between economic responsibilities and other responsibilities such as legal and ethical responsibilities. It prioritises economic utility, not sustainable development (Heikkurinen and Bonnedahl, 2013).

The alternative line of argument regarding a company's behaviour is its modern form which targets all three sustainability pillars, more often called business lines: the economic, environmental (e.g., natural) and social lines. Based on this understanding, CSR is now understood as a business' contribution to corporate sustainability (van Marrewijk, 2003). This corporate understanding corresponds to the societal approach to CSR which recognises that companies are an integral part of society and thus responsible to society as a whole. It assumes that the company considers a broader set of stakeholders (not only its customers) in its behaviour. The ethical behaviour of a firm reflects the concerns of its consumers, employees, shareholders and the community (Caroll, 1991). The stakeholder orientation focuses on the broader responsibility of a firm through its stakeholders who affect the business, thus incorporating all three sustainability pillars. It does not draw a strict line between the firm's economic responsibilities and other responsibilities like ethical responsibilities.

Further, the firm's reluctance to talk about sustainability is illustrated by another practice example of John Elkington who started his UK-based private consulting company, SustAinable. He invented a new terminology to sell sustainability business practices to corporate and private business. He named this new approach the *triple bottom line* (TBL) (Elkington, 1997). The TBL is just another way of wording the three pillars of sustainability, yet it is more strongly accepted by corporate stakeholders. The TBL concept has penetrated some of the tourism debate, but much less so than the sustainability concept.

A similar argument can be made for the notion of *green tourism* which organisations have in many cases used as a synonym for sustainable tourism. More specifically, the British Green Tourism Business Scheme claims "Green tourism is sustainable tourism ..." (GT, 2014). The same sentiment can be found on the web pages of other national tourism organisations as the term green tourism is often more accepted by different stakeholder groups than that of sustainable tourism (Forquan et al., 2010 and STO., 2013). For example, after trying to promote sustainable tourism for many years with a very limited response from tourism business sector, Slovenian national tourism organisation strategically introduced *green tourism* as a synonym for sustainable tourism in 2009 and succeed to mobilise diverse tourism stakeholders to join the efforts under the slogan "Slovenia is green, Slovenia acts green, Slovenia promotes green" (Mihalič, 2013 and STO., 2013), targeting all three sustainability pillars.

From a modern perspective, the debate on the differences between sustainable, CSR, TBL or green tourism seems unproductive because all converge and relate to the same pillars. However, from the aspect of business and global market reality, which delineates the roles of business, governments and civil society, the separated classical shareholders business approach still needs attention. The CSR or TBL notion seems to be more strongly rooted in the world of business corporations, and the notion of sustainability in the public sector and sustainable tourism destination development. This fact alone might be evidence of the gap between the differences presented above between an orientation towards shareholder profit and one towards wider societal responsibility and is supported in the context of the above separation.

Although there is some evidence that responsibility concepts in from of CSR models are applied to private and public sector tourism practice, the notion of responsible tourism has evolved separately. It has been present in tourism in some form since the early 1980s and is connected to many studies on environmental problems in tourism (Bramwell et al., 2008, Butler, 1995, Chettiparamb and Kokkranikal, 2012, Goodwin, 2011, Leslie, 2012b and Leslie, 2012c). Contrary to the relative consensus on the three-pillar meaning of sustainable tourism, observed as a concept, the notion of responsible tourism (or CSR) has not been as strongly established or as frequently applied.

According to some authors, responsibility and responsible tourism can mean "anything" (Chettiparamb and Kokkranikal, 2012, p. 302; van Marrewijk, 2003), and its use adds nothing to the conceptual understanding of tourism. The existing literature and documents on responsible tourism reveal many understandings that: vary regarding whether the subject is the concept, theory or practice (Leslie, 2012a, Leslie, 2012b and Leslie, 2012c); involve different stakeholders and different aims; refer to social or polar tourism (ICRT, 2011); or create environmentally friendlier tourism or better places for people to live and visit (RTP, 2002). Further, the existing literature focuses on responsible consumption or production (Budeanu, 2005 and Stanford, 2008), explores actor relations such as the tourist-host population, and questions individual versus social responsibility as well as the political assumptions and governance models behind the responsibility (Bramwell et al., 2008 and Hall, 2012).

In addition to these quite diverse research interests, it seems that the use of *responsible tourism* often occurs when Krippendorf's legacy (1987) of a new understanding of tourism is emphasised (Bramwell et al., 2008 and Goodwin, 2011) and when questions of ethical and moral responsibility are brought to the fore (Blackstock et al., 2008, Bramwell et al., 2008 and Fennell, 2006). Moreover, it occurs when the *responsibility* of tourism needs to be stressed as some implementation difficulties have been identified and a sustainability agenda is being created. In this regard, we claim that the notion of *responsibility* relates to responsible *behaviour and action*.

This behaviour- or action-based view of responsible tourism has been supported by the work of many researchers. The guru of the responsible tourism movement, HaroldGoodwin (2011), claims that:

"The idea of Responsible Tourism has at its core the imperative to take responsibility, to take action; consumers, suppliers and governments all have responsibilities. The ambition of Responsible tourism is to address the impacts of mainstream tourism, to enhance the positive and to reduce the negative ... (p. 2) .... *Responsible Tourism is about everyone involved taking responsibility for making tourism more sustainable"* (p. 31).

Goodwin clearly links responsible tourism to action in favour of making tourism more sustainable. His understanding of the concept of responsibility assumes three aspects: accountability, capacity to act, and the capacity to respond. It is the third aspect, response-ability that directly addresses tourism behaviour and involves entering into a dialogue, creating solutions and acting to make tourism more sustainable.

This behaviour-based view has also been supported by David Leslie (2012b). In his meaningfully titled book chapter, "The Responsible Tourism Debate", he argues that responsible tourism is *"a behavioural trait ... based on the basic principles of respect for others and their environment..."* (p. 20).

This view of responsible tourism enabled Leslie to go further and note that responsible tourism assumes "acting responsibly in terms of one's own actions, and moreover, in the management and operation of business" (p. 20). When faced with the dilemma of a conceptual basis for judging responsibility, he refers to the ethical and environmental criticisms of tourism and suggests behaving "... environmentally and/or ethically responsible" (p. 20).

A step further has been taken in the research by Chettiparamb and Kokkranikal (2012), meaningfully entitled "Responsible Tourism and Sustainability". The authors also conceive of responsible tourism as behaviour-based and argue that the notion is linked to sustainability initiatives, which gives a solid conceptual substance to their understanding of responsible tourism. In their research, they progress to the CSR concept and discuss a tourism-specific manifestation of the private sector CSR agenda, which focuses on *"... facilitating responsible behaviour ... through CSR-related practices targeting wider socio-economic issues"* (Chettiparamb and Kokkranikal, 2012, p. 307).

Facilitating responsible behaviour creates many questions. There are questions of whether the environmental concern of the tourism industry results in more responsible behaviour and which practical measures and policies encourage and enable greater responsibility (Bramwell et al., 2008 and Goodwin and Francis, 2003). Another important question is who the relevant actors are, who should make the relevant choices and what type of philosophy and motivation underpins the responsible tourism paradigm (Fennell and Malloy, 2007). They are all relevant to our tourism discourse as each adds a piece of evidence from a different perspective, demonstrating the discourse's complexity and differentiation, demanding that fragmented issues be placed within a common framework.

Further, many action-based tourism documents can be listed to illustrate the more practice-based use of the notion of responsible tourism. Although the term responsible tourism appeared in many older documents, the first declaration on responsible tourism was published in 2002 ("Cape Town Declaration on Responsible Tourism," 2002). In 2012, the UNWTO Ethics Committee decided to introduce and apply the term *responsible tourism* as part of the new logo for the "Global Code of Ethics for Tourism" (Diotallevi, 2013). Strongly focusing on sustainable development and adopted at the WTO General Assembly in 1999, the Code was given a new cover page and the subtitle "For Responsible Tourism". It is also expected that the UNWTO Secretariat and Committee will work on proposing an official definition of responsible tourism (Diotallevi, 2013).

In addition, the new "European Charter for Sustainable and Responsible Tourism" (TSG, 2012) supports our argument. The charter is built on an old Tourism Sustainability Group (TSG) document (TSG, 2007) which dealt exclusively with the notion of sustainable tourism ("Action for More Sustainable European Tourism"). However, five years later, the new document uses the terms sustainable and responsible tourism but omits the word action. "Action" is now incorporated in the broader understanding of the notion of responsible tourism, as the latter is defined as:

awareness, decisions and actions of all those involved in the planning, delivery and consumption of tourism, so that it is sustainable over time (TSG, 2012: 2).

Bearing this definition and the above discussion in mind, responsible tourism is not a synonym for sustainable tourism. Responsible tourism addresses the aforementioned sustainable tourism discourse in implementation and is more of an expression to describe tourism that is sustainable because it acts sustainably. It is nothing new, neither a new alternative to other tourism forms nor a new way of "doing" sustainable tourism, as some authors claim (Blackstock et al., 2008, p. 279). Thus, responsible tourism builds on appropriate sustainability-based strategies and policies and adds appropriate behaviour, meaning sustainable (re)actions or, according to Goodwin's concept, response-ability, underpinned by tourism environmental awareness and ethics (Fennell, 2006). This sustainability–responsibility connection is illustrated in Fig. 1 in which the pillars of sustainable development stand upon the sustainable tourism requirements previously presented in the section of the paper on the "Sustainable-responsible Tourism Discourse". Indeed, the roof of the Fig. 1 structure should be termed *responsustable tourism*, demonstrating both responsibility and sustainability in tourism behaviour.

Of course, to answer the question regarding to what extent sustainability principles have improved the real tourism sector's responsustability, tourism sustainability indicators might be of assistance. In tourism research and applied practice, many models can be found, and their use is quite popular. Indeed, sustainable tourism indicators can be observed as tourism responsibility measures and monitoring that support environmentally friendlier tourism management and practice (Blackstock et al., 2008). However, the sustainability indicator schemes are very heavily measurement- and thereby result-oriented, and the majority fail to understand that achieving sustainability in a sector should be understood as a social transformation process and that old sustainability indicator vocabulary might not be sufficient at this point in time. This paper attempts to add to this perspective.

4. MODELLING RESPONSUSTABLE TOURISM

4.1. Environmental damage theories

Returning to our initial discussion, which addressed our concern about environmental damage and tourism, the question of the occurrence of environmental damage becomes relevant. Many researchers have tried to explain why "Our Common Future" awareness is not incorporated into our current behaviour. According to these scholars, the answer is threefold (Frey, 1985 and Mihalič and Kaspar, 1996). Environmentally improper behaviour is first observed as a result of the non-functionality of the global laissez-fair capitalism market system and as related to the problems of public goods, property rights and externalities. Second, the same problem can be explained based on economic and population growth, which results in overproduction and the overconsumption of resources. The final explanation blames improper behaviour based on human ignorance and/or the absence of environmental awareness and ethics.

All of the above theories are relevant to our discourse. Regarding the first theory, the environmental goods are wrongly treated as a public goods, as they don't have an owner, are assumed to be plenty and are non-excludable in their consumption. Consequently, these goods are used (and overused and polluted) freely regardless of a user's participation in the related costs. Theory sees the problem in environmental goods, being external to the market allocation mechanism which has the ability to determine the price, through the confrontation of demand (willingness to pay) and supply (willingness to sell). Thus environmental, e.g. external goods must be internalised to our current neo-liberal market allocation system in order to be properly allocated. In theory, such a system would guarantee the socially optimal consumption and use of environmental goods. However, a lead tourism researcher in the criticism of market-led tourism systems,

Freya Higgins-Desbiolles, 2006, Higgins-Desbiolles, 2008 and Higgins-Desbiolles, 2010, argues that the current system of market-based neoliberalism and its culture and ideology of consumerism are inherently unsustainable and calls for a less damaging system that will enable tourism to be a positive developmental social force. Instead of a private-sector commercial profit-driven corporatized tourism paradigm, she suggests an alternative paradigm based on equity, justice and sustainability that is civil-society- and government-driven.

The second theory on environmental damage, which addresses the questions of population and economic growth, is relevant to tourism due to tourism demand and supply growth. It is ever present in the tourism literature and is referred to in almost every work that relates to tourism development and business. This is the debate on sea, sun and sand mass tourism and its seasonal concentration (Aguiló et al., 2005, Figini and Vici, 2012, Fink, 1970, Mihalič, 2006 and Poon, 1993), which has often been observed as a dominant cause of the current sustainability tourism discourse. In this regard, non-mass-tourism developmental models, e.g., alternative tourism, are called for. At the same time, it is argued that because sustainable tourism is a set of principles, not a type of tourism, it can be applied to any tourism type or destination type, including concentrated mass tourism destinations (UNEP & WTO, 2005).

However, it is the third theory on the occurrence of environmental damage that is the most relevant to this paper. Why does tourism not behave responsibly towards Our Common Future, and why does it not develop and function in a sustainable way? On one hand, tourism businesses that are only driven by profit motives and ignore environmental and social awareness, ethics and policy in terms of strategy-based actions will necessarily face a sustainability discourse. However, on the other hand, economic business thinking is based on rationality and economic values and perceives ethical demands for ecology and justice as irrational. Based on this argument, the theory claims that the absence of so-called social environmental ethics has caused the current negative attitude to our environment (Frey, 1985). In this regard, the existence of environmental damage is explained through the absence of environmental social ethics (and awareness) and as a product of human ignorance. Thus, application of this theory calls for increased environmental awareness and ethics. It should, however, be noted that the theory assumes that people are good by nature and that a proper course of action is followed once they have enough information and knowledge on proper and just behaviour, regardless of the current economic allocation system in place.

4.2. Social stages of responsible behaviour

The information, knowledge, awareness and ethical aspects have been further addressed by the model of the penetration of environmental awareness into society and its policy. To explain how these aspects are addressed, we can turn to political economy. The Swiss-born welfare economist Bruno Frey (1985) proposed the famous four-stage scheme for new environmental trends: *Stage 1: pre-environmental-protection thinking*(society is simply unaware of environmental problems); *Stage 2: awareness of environmental harm* (which explodes after a society can no longer deny the facts); *Stage 3: discussion about environmental policy* (with the intention to mobilise for proper action); and *Stage 4: implementation of environmental policy* (which involves carrying out the appropriate action and assumes proper results).

In the continuation, Frey discussed the above scheme in relation to different stakeholders such as the voters, government, bureaucracy and producers (Frey, 1992). Although this is also highly relevant to our tourism-related debate, this paper only focuses on the application of Frey's model to the tourism sustainability discourse.

4.3. The Triple-A Model on sustainable-responsible tourism: towards responsustable tourism

One can easily apply the same four phases to the way our tourism stakeholders address environmental concern and sustainability. The idea of Frey's social stages was first applied to the tourism field by Mihalič and Kaspar (1996). Here the original four stages were reinvented and renamed. The first stage is the so-called *ignorance* stage that occurs before the destination considers environmental problems. Tourism stakeholders are unaware of such problems and thus express no concern at all; other, non-environmental sets of values determine their behaviour. As this phase acts somewhat as the pre-history of our sustainability discourse, the next stage is more relevant. This is a stage of environmental *awareness*, as exemplified by exposure to information on tourism environmental impacts and relationships that have made destinations sufficiently aware of tourism impacts that they can no longer be denied. Since tourism environmental concern, our starting point, is deeply connected with the current sustainable tourism ideology, sustainability issues inevitably enter the destination and create and shape so-called tourism sustainability awareness. The next stage involves the conceptual inclusion of sustainability issues, as defined above, because these become a matter of debate and are transferred into goals, coded into the destination's strategy and placed on its *agenda*. To shape and form the willingness to act consistently with the sustainability concept, the discussion of relevant policy instruments begins. Finally, the social-sustainability ethics and the need to act to achieve the strategic goals drive this process toward and into the last stage, which is the successful implementation of a sustainability policy and, hence, the stage of implementing sustainable *action*, thereby manifesting environmental responsibility. The destination, e.g., its stakeholders, behave in a sustainable manner, thus meeting the objectives and standards for sustainable development, business, products, etc.

The crucial words here are *Awareness–Agenda–Action*, all starting with the letter "A" and thus creating our Triple-A Model. Henceforth, we refer to this model as our Triple-A Model, which is presented and explained in Fig. 2.

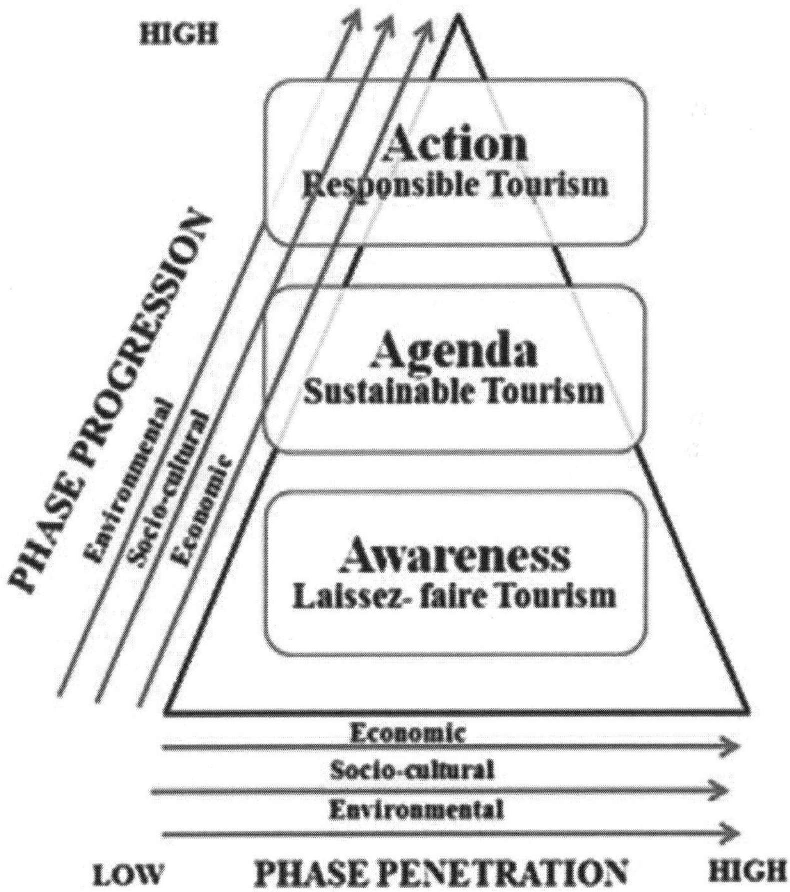

Figure 2. The Triple-A Model – for responsustable tourism.

Fig. 2 presents the top three stages of Frey's model after environmental concern has entered society. The pyramidal form associates the sustainability–responsibility process with mountain climbing. To reach the top of a mountain, a mountain climber needs to start at the bottom. He knows that the stepwise passage through the lower mountain station is a precondition for reaching the higher levels and that each climbing expedition up needs to be supported by a base station below. Thus, the narrowing shape of the pyramid suggests that each previous stage is a base upon which the next stage is dependent. The vertical and horizontal arrows also demonstrate that the penetration of sustainability into a destination is a twofold process. First, the process moves stepwise because the destination has to pass through different phases – from the starting phase of the destination's awareness, through the agenda stage to achieve the implementation action phase at the top of the pyramid. This process is demonstrated by the vertical progression line. Second, the process is continuous because the destination may proceed to the next phase before the previous phase has fully matured. More specifically, the top stage might be reached by only a few of the mountain expedition's climbers; the rest might stay in lower base stations and climb up later or not at all. This characteristic is presented by a horizontal arrow line that measures the penetration into each stage. Theoretically, a destination and its tourism stakeholders might find themselves at any of the stages on the sustainability–responsibility progression and penetration scale. At the same time, the destination might reach the top phase without having fully developed the characteristics of the previous phases, thus simultaneously positioning its stakeholders in all phases.

The challenge of our pyramid is that it is based on a three-pillar sustainability understanding and thus reflects awareness, agenda and action in three different areas: economic, environmental and socio-cultural. This is illustrated by the three penetration and progression arrows in Fig. 2. In fact, the pyramid in Fig. 2 is the sum of three different pyramids and can be broken down into those three, one for each pillar. Tourism stakeholders might be at different positions in terms of both progression and penetration in each of the three pyramids. The existing global market system prioritises progression in the economic pillar pyramid; for some stakeholders, the other two pyramids (environmental and social) might become less important, thus affecting the triple sustainability of the complete pyramid. For example, one hotel study that measured firm behaviour with regard to the presence of different strategies showed that most of the surveyed hotels have an economic strategy in place (reported for 80% of firms) and only 35% of firms confirmed they had an all-pillars sustainable development strategy (Mihalič et al., 2012).

In reality, many destinations and tourism businesses do not know exactly where on this pyramidal scale they are and how sufficiently or insufficiently they have developed each of the phases. For example, a destination or firm might be well informed about the occurrence of environmental damage and be aware of environmental problems but might lack the appropriate information and knowledge on how to respond to those problems. The greater the information gap, the further to the left it will be positioned on the phase penetration scale. Some destinations or firms might have recently discussed sustainability agendas and created partnerships to adopt one, while others might have already included sustainability in a strategy and are confronted with policy-efficacy or implementation problems because a destination's stakeholder behaviour might lag behind the declared sustainability goals, meaning that the critical mass has not conquered the apex of the sustainability mountain.

5. RESULTS AND DISCUSSION: A FRAMEWORK FOR UNDERSTANDING THE SUSTAINABLE-RESPONSIBLE TOURISM DISCOURSE

Table 1 links our understanding of responsustable tourism presented in Fig. 1 with our Triple-A Model presented in Fig. 2. Further, it demonstrates how the whole process is driven by visitor satisfaction and their will to pay for sustainability, an ultimate measure and driver of a destination's success.

Column 4 of Table 1 employs the term *environmental laissez-faire tourism* to describe a tourism realm that is free of any environmentally based interference, including values, enabling its system to operate according to economic rules and only one (economic) business line. Tourism is free of any environmental demands or standards, such as legal interventions, public policy, voluntary environmental actions or social mobilisation and other non-economic instruments and demands. As long as tourism is only based on the sustainable tourism concept, the sector still behaves in much the same way as before. Practically speaking, responsible tourism behaviour only begins in the third stage of our tourism model, when a sustainability-based agenda for responsible tourism starts to be implemented.

5. Results and discussion: a framework for understanding the sustainable-responsible tourism discourse

Table 1. A framework for understanding the sustainable-responsible tourism discourse.

Frey's model	Responsustable tourism pillars and requirements (Fig. 1)	The Triple-A Model (Fig. 2)	Stage-related tourism concept	Description
1	2	3	4	5
4. Implementation of environmental policy	Stakeholders' participation, cooperation, consensus, leadership, action & Visitor satisfaction	**3. Action**	Responsible tourism behaviour	The process of engaging in actual behaviour to achieve a full set of sustainability aims (responsible behaviour), taking actions in line with all three pillars of the sustainability concept, implementation of the agenda.
3. Discussion of environmental policy	Economic and/or Environmental and/or socio-cultural pillar & Visitor satisfaction	**2. Agenda**	Sustainable tourism concept	A blueprint for action for more sustainable tourism. The agenda is based on a sustainable tourism strategy and informed by sustainability value sets that demand action to implement the sustainable tourism concept. Economic considerations are extended to environmental and socio-cultural aspects. All are observed as important pillars.
2. Awareness of environmental harm	Tourism environmental awareness, education, information, ethics & Visitor satisfaction	**1. Awareness**	Environmentally laissez-faire tourism	Despite growing concern about tourism and its effects on the social, cultural and natural environment, the tourism sector is fully led by market rules, transactions among private parties are free from all non-economic considerations.
1. Pre-environmental-protection thinking		Ignorance(*Not presented in our Tripple-A Model in* Fig. 1)	Environmentally laissez-faire tourism	A pre-historic stage of our sustainable-responsible tourism discourse, the tourism sector is unaware of socio-cultural and environmental tourism issues.

One may argue that this paper is dwelling on old findings because the concept-action tourism discourse has been present in the sustainability debate for quite some time. For example, the original Brundtland Report offered "... the recommendations for a sustainable course of development" (WCED, 1987: 16). Indeed, the Agenda 21 approach used to be quite popular, referring to a blueprint for action for more sustainable real tourism. However, this approach has not reduced our sustainability discourse because*agenda* is our final Alpine base camp and thus serves as the second A in our Triple-A Model. *Agenda* is a precondition for our mountain climber to start their journey to the peak of the mountain, to start the implementation action phase. Agenda extends a theoretical sustainability concept into a concept with specifically defined actions, meaning that it codifies the sustainability principles into a strategy and policy and offers policy instruments. Thus, tourism that enters the implementation action stage needs a new name. Based on the proper action or responsible actual behaviour, the term *responsible tourism* would be appropriate. This interpretation argues that responsible tourism is not a new tourism concept and especially not an alternative concept to sustainable tourism. It simply refers to acting responsibly and in line with sustainability values. Thus, responsible tourism is a result of the transition towards actual tourism based on more sustainable values.

The Triple-A Model pyramidal view of the progression from environmentally laissez-faire tourism towards responsible and thus more sustainable tourism integrates new parts that have developed within the tourism environmental and sustainability debate. It demonstrates an increase in the interconnectedness and complexity of the debate, integrating its stages, which include and transcend the previous ones. However, vertical and horizontal progression up our Awareness–Agenda–Action pyramid is not guaranteed or self-evident; however, as time progresses, it cannot be stopped. For example, the awareness- and ethics-driven transition towards responsible tourism might be slowed down for many reasons. One study found that European tour operators have a high level of awareness about tourism's negative impacts and recognised their responsibility to reduce them. Nevertheless, facing low customer demand for responsible tourism practices and having little regulatory pressure to behave in a more environmentally friendly manner, tour operators engage in few concrete activities and these remain superficial relative to their potential due to their central distribution role and capability to direct customers to suppliers and destinations (Budeanu, 2005 and Sigala, 2008). However, tour operators are financing sustainability and environmental research and customer education activities, which will straighten the vertical position of tourism in our awareness stage and allow more stakeholders to progress to the next stage in the pyramid where more activities will be demanded and designed in both the private and public sectors. This case clearly demonstrates the shareholder market-led approach to corporate sustainability. However, ignoring the progression in time might be dangerous as standards for environmental behaviour are nevertheless changing due to the above-mentioned processes. In other words, behaviour is changing because "... the way the world is going, the companies are having to become more responsible ... it ... helps them to remain competitive, so they are going to have to do it" (Miller, 2001, p. 595).

Our Awareness–Agenda–Action understanding of tourism responsibility is in full harmony with the European Commission's latest standing on sustainable versus responsible tourism (Mihalič, 2011), which demands "*... awareness, decisions and actions ..., so that it is sustainable over time*" (TSG, 2012, p. 2). However, this explains little about the corporate and economic tensions underlying it.

These tensions might be best illustrated by the confrontation of two popular concepts: sustainability versus CSR. In this context, the question of the similarities and differences between sustainability and CSR may arise. This discussion can take place at the conceptual level where both sustainability and CSR are observed as two concepts that converged over time and are now both linked to the same three pillars: economic, social and environmental. However, it is evident that CRS is more popular in the corporate world and sustainability among tourism destinations and public bodies or organisations. One may even claim that the term sustainability has been avoided by corporate business practice and consultants.

We argue here that CSR, which is relevant to the corporate world, operates from the standpoint of corporate values and may add the other two sets of sustainability values, e.g., social and environmental, in a way that makes real business or economic sense. At the profit-driven level of sustainability motivation (market-oriented view), sustainability is narrowly defined inside the business fence and is only promoted if profitable. Quite differently, debate based on sustainable tourism (broader stakeholder orientation) claims that all pillars are equally important, with no priorities given to any pillar and certainly not to the economic one. In this case of a holistic level of sustainability, the motivation for a sustainability orientation is derived from the awareness that all beings and phenomena are mutually interdependent, and universal responsibility towards all other beings is fully recognised (van Marrewijk, 2003). As research has shown, in the laissez-faire environmental tourism phase, tourism managers publicly admit economic but not ecological responsibility, believing that they are a priori contradictory. After society enters the awareness stage, where the connections between the economy, society and ecology become obvious and cannot be ignored, such statements become rare and many managers participate in the discussion on the sustainability agenda. However, the question of how to convince competitive corporations to follow the full set of sustainability values remains unsettled. Ultimately, the idea of sustainability has emerged completely isolated from a thorough analysis of the principles governing market-based economic systems, and the corporate world seems to have a recognised right to acknowledge and protect the importance of economic values and even challenge the other two pillars. In this regard, sustainability seems to live in an isolated world and CSR (if applied) in a so-called green capitalism characterised by "green capitalists" and "green consumers" (Elkington, 1997, pp. ix, x). Quite the contrary, tourism destinations are expected to behave in a sustainable-responsible fashion and a destination's responsibility is broadly defined inside and outside the destination's fence. The question here is whether the all-pillars sustainability position is observed as a threat to businesses and profit-driven sustainability as a legitimate mundus operandi in line with corporate values and practice. The obvious conclusion is that a CRS orientation assumes that tourism can happily live in "sustainable capitalism", underpinned by green customers who are demanding more green products and thus more responsible tourism, and are willing to pay for it.

To summarise the above discussion, there are two reasons for the rise of *responsible tourism*. The first is the need to overcome what has become evident, the actual sustainability discourse, due to the inability of sustainable tourism itself to be more strongly implemented in real tourism. The second reason is more ideology-based and refers to an attempt to merge the tourism-based environmental debate with the language used in other sectors, most importantly the corporate or business world, with the aim of increasing the acceptance and attractiveness of the relevant tourism concept among all tourism stakeholders.

6. CONCLUSIONS AND FURTHER RESEARCH

The low efficacy of *sustainability* in tourism coupled with the increasing attention to the notion of *responsible tourism*, as demonstrated by current theoretical and practical developments, call for critical reflection. The changes in our understanding occur in stages that follow both a logical and a chronological order. More holistic understanding develops later because it has to wait for the emergence of the parts that will then make up a complete theory and interconnections with the new stages including and transcending the previous ones (van Marrewijk, 2003).

In this context, the present conceptual paper integrates and challenges different existing theoretical and practical private and public stakeholder understandings and uses of the notions of responsible and sustainable tourism. It offers a well-reasoned integrated understanding of responsible tourism and the process that leads towards *responsustable tourism*. The proposed Triple-A Model helps to analyse and diagnose the main sustainability tourism discourse on enduring unsustainable tourism practices. The creation of the model was also provoked by the current inconsistency in use of the terms sustainable and responsible tourism and theories on environmental damage and behavioural reaction to tourism damage. As the two developments, theoretical and practical, are relevant to this paper, we focussed our discussion on both.

By putting the sustainability–responsibility discourse into a relevant context of political economics and behavioural economics, this paper adds to tourism concepts and nomenclatures. Although the proposed Triple-A Model and its phases have existed in some form in the tourism debate and while one might argue that they are common sense, they contribute to understanding the process and the reality of the sustainable tourism discourse. This understanding, placed in a real context, brings us to the tourism-responsibility paradox. Although the Awareness-Agenda-Action process theoretically leads towards more socially and environmentally responsible tourism, it is the economic set of interests that is attempting to react and adapt to the growing environmental awareness and demand and is leading the process and promoting the notion of responsible tourism. Here, we refer to our observation that its use is pushed by the 'responsibility' (e.g., in the form of CRS) debate, recognising the need to go green but while strongly protecting the values of the economic pillar. Thus, more research is needed to assess whether the above-mentioned green movement determined by the capitalistic system is in fact leading towards a sufficiently sustainable future.

Further, our model can help increase tourism responsibility by analysing the implementation chances of agendas and environmental policies. To avoid the expectation that the model leads to the implementation (e.g., responsible tourism) phase itself, the model must recognise that potential conflicts among the stakeholder groups are an important element in the political process, enabling progression towards the higher phases. In that regard, the roles of tourists, governments, the tourism industry, NGOs and local populations will also need to be further discussed and researched.

Our model can be applied to tourism destinations of any level, tourism businesses, governments, non-governmental organisations, institutions and tourists. All of these relevant stakeholders have a responsibility to seek to reduce the negative and enhance the positive environmental, socio-cultural and economic impacts of tourism. In this regard, more research is needed from the aspect of each group, focusing on awareness (and ethics) creation, instruments for achieving the goals and behavioural context.

One may extend the understanding of responsible tourism behaviour to all stages of our pyramid. That is, if one understands that the destination is responsible for being aware of tourism damage and minimising it, that responsibility also requires the developing of an action plan. Thus, responsible tourism might refer to all of the stages, not only the last one, as presented in Fig. 2. Indeed, we used the term responsible tourism in a narrower sense, limited to actual behaviour that makes real tourism more sustainable. In our model, we have connected responsible behaviour with the sustainable tourism concept and thus added the sustainability value to the behavioural concept. In this context, another, joint nomination would better describe our top pyramidal tourism phase. Merging the words responsible (behaviour-based) and sustainable (concept and values-based) produces the new term responsustable tourism. It is argued that this new term fully reflects the academic and practical debate and action that is increasingly labelled "responsible" tourism, yet de facto based on sustainability. In this context, paper proposes a Triple-A Model as a tool that helps to understand the process of how a responsible tourism destination or firm actually implements sustainability agenda.

REFERENCES

1. Aguilo, E., Alegre, J.n., Sard, M., 2005. The persistence of the sun and sand tourism model. Tour. Manag. 26 (2), 219–231.
2. Blackstock, K.L., White, V., McCrum, G., Scott, A., Hunter, C., 2008. Measuring responsibility: an appraisal of a Scottish National Park's sustainable tourism indicators. J. Sustain. Tour. 16 (3), 276–297.
3. Bohdanowicz, P., Simanic, B., Martinac, I.V.O., 2005. Environmental training and measures at scandic hotels, Sweden. Tour. Rev. Int. 9 (1), 7–19.
4. Bramwell, B., Lane, B., 1993. Sustainable tourism: an evolving global approach. J. Sustain. Tour. 1 (1), 1–5.
5. Bramwell, B., Lane, B., McCabe, S., Mosedale, J., Scarles, C., 2008. Research perspectives on responsible tourism. J. Sustain. Tour. 16 (3), 253e257.
6. Budeanu, A., 2005. Impacts and responsibilities for sustainable tourism: a tour operator's perspective. J. Clean. Prod. 13 (2), 89–97.

7. Butler, R., 1995. Introduction. In: Butler, R., Pearce, D. (Eds.), Change in Tourism: People, Places and Processes. Routledge, New York, pp. 1–11.

8. Caroll, A.B., 1991. The Pyramid of Corporate Social Responsibility: toward the Moral Management of Organizational Stakeholders Business Horizons. July – August.

9. Chettiparamb, A., Kokkranikal, J., 2012. Responsible tourism and sustainability: the case of Kumarakom in Kerala, India. J. Policy Res. Tour. Leis. Events 4 (3), 302–326.

10. RTD7 Conference: Responsible Tourism in Destinations, 1–4 October 2013. Catalunya, Barcelona. Retrieved 1.12.2013, from: http://www. accessibletourism.org/ ?i¼enat.en.events.1473.

11. Dasmann, R.F., Freeman, P.H., Milton, J.P., 1973. Ecological Principles for Economic Development. John Wiley & Sons, Washington.

12. Diotallevi, M., 2013. Personal Correspondence, 14.3.2013, Chief of Programme Ethics $ Social Dimensions of Tourism. UNW, Madrid.

13. Dwyer, L., 2005. Relevance of triple bottom line reporting to achievement of sustainable tourism: a scoping study. Tour. Rev. Int. 9 (1), 79–938.

14. EC, 2014. European Tourism Indicators System Toolkit for Sustainable Destinations. European Commission, Brussels.

15. ECETAT, ECOTRANS, 2004. The Visit Initiative, Tourism Eco Labelling in Europe e Moving the Markets towards Sustainability. European Centre for Eco Agro Tourism, Amsterdam.

16. Elkington, J., 1997. Cannibals with Forks: the Triple Bottom Line of 21st Century Business. Capstone, Oxford.

17. Fennell, D.A., 2006. Tourism Ethics. Channel View, Clevedon.

18. Fennell, D., Malloy, D., 2007. Codes of Ethics in Tourism. Practice, Theory, Synthesis. Channel View Publications, Clevedon.

19. Figini, P., Vici, L., 2012. Off-season tourists and the cultural offer of a mass-tourism destination: the case of Rimini. Tour. Manag. 33 (4), 825–839.

20. Fink, C., 1970. Die Massentourismus (Mass Tourism). Verlag Paul Haupt und Stuttgart.

21. Forquan, A.M., A., P., Hussin, R., 2010. Promoting Green tourism for future sustainability. Theor. Empir. Res. Urban Manag. 5 (8), 64–74.

22. Frey, B.S., 1985. Umweltoekonomie (Engl.: Environmental Economics). Vandenhoeck & Ruprecht in Goettingen, Goettingen.

23. Frey, B.S., 1992. Umweltokonomie (Engl.: Environmental Economics). Vandenhoeck € & Ruprecht, Gottingen €.

24. Goodwin, H., 2011. Taking Responsibility for Tourism. Goodfellow Publishers Limited, Oxford.

25. Goodwin, H., Francis, J., 2003. Ethical and responsible tourism: consumer trends in the UK. J. Vacat. Mark. 9 (3), 271–284.

26. GT, 2014. Why Green Tourism? Green Tourism. Retrieved 15.4.2014 from: http:// www.green-tourism.com/.

27. Hall, M., 2012. Governance and responsible tourism. In: Leslie, D. (Ed.), Responsible Tourism. Concepts, Theory and Practice. CABI, Wallingford, pp. 107–118.

28. Heikkurinen, P., Bonnedahl, K.J., 2013. Corporate responsibility for sustainable development: a review and conceptual comparison of market- and stakeholderoriented strategies. J. Clean. Prod. 43 (0), 191–198.

29. Higgins-Desbiolles, F., 2006. More than an "industry": the forgotten power of tourism as a social force. Tour. Manag. 27 (6), 1192–1208.

30. Higgins-Desbiolles, F., 2008. Justice tourism and alternative globalisation. J. Sustain. Tour. 16 (3), 345–364.

31. Higgins-Desbiolles, F., 2010. The elusiveness of sustainability in tourism: the culture-ideology of consumerism and its implications. Tour. Hosp. Res. 10 (2), 116–115.

32. ICRT, 2011. Alberta declaration on responsible tourism in destinations. In: 5th International Conference on Responsible Tourism in Destinations. Canada.

33. Inskeep, E., 1991. Tourism Planning: an Integrated and Sustainable Development Approach. John Wiley & Sons, New York.

34. International Conference: Sustainability Issues and Challenges in Tourism, 3e5 October 2013. Retrieved 1.12.2013 from. http://www. butrmconferences.org/.

35. IUCN, 1980. World Conservation Strategy: Living Resource Conservation for Sustainable Development. International Union for Conservation of Nature and Natural Resources, Gland.

36. Kaspar, C., 1973. Fremderverkehrsoekologie e eine neue Dimension der Fremdenverkehslehre. (Engl: tourism ecology e a

new dimension of tourism sience). In: Walter, E. (Ed.), Festscrift zur Vollendung des 65. Lebensjahres von P. Bernecker, pp. 139–143. Wien.

37. Krippendorf, J., 1984. Die Ferienmenschen. (Engl.: Holiday People). Orel Fuessli Verlag, Zuerich.

38. Krippendorf, J., 1987. The Holiday Makers e Understanding the Impact of Leisure and Travel. Heinemann, Oxford.

39. Leslie, D., 2012a. Responsible Tourism. Concepts, Theory and Practice. CABI, Wallingford.

40. Leslie, D., 2012b. Introduction. In: Leslie, D. (Ed.), Responsible Tourism. Concepts, Theory and Practice. CABI, Wallington, pp. 1–16.

41. Leslie, D., 2012c. The responsible tourism debate. In: Leslie, D. (Ed.), Responsible Tourism. Concepts, Theory and Practice. CABI, Wallingford, pp. 17–42.

42. McGugan, S., 2002. Asynchronous computer mediated conferencing to support learning and teaching: an action research approach. J. Hosp. Leis. Sport Tour. 1 (1), 29–42.

43. Mihalic, T., 2006. Tourism and its Environments: Ecological, Economic and Political Sustainability Issues. Ekonomska fakulteta. Univerza v Ljubljani, Ljubljana.

44. Mihalic, T., 2011. Discussion on Sustainability, Responsibility and CSR, Based on the TSG Working Document. TSG meeting 20.6.2011. European Commission, Brussels.

45. Mihalic, T., 2013. A Green tourism Barometer in the time of economic crisis e the concept and case of Slovenia. In: Paper presented at the conference Tourism in Southern and Eastern Europe. Retrieved 1.12., 2013, from. http://www.fthm.uniri.hr/files/Kongresi/ ToSEE/2013/ papers/1%20Mihalic.pdf.

46. Mihalic, T., Kaspar, C., 1996. Umweltokonomie im Tourismus. (Engl.: Environmental € Economics in Tourism.). Paul Haupt, Bern.

47. Mihalic, T., Zabkar, V., Knezevic Cvelbar, L., 2012. Hotel sustainability business model: evidence from Slovenia. J. Sustain. Tour. 20 (5), 701–719.

48. Miller, G., 2001. Corporate responsibility in the UK tourism industry. Tour. Manag. 22 (6), 589–598.

49. Poon, A., 1989. Consumer strategies for a new tourism. In: Cooper, C. (Ed.), Progress in Tourism, Recreation and Hospitality Management, vol. 1. Bellhaven Press, London, pp. 91–102.

50. Poon, A., 1993. Tourism, Technology and Competitive Strategies. CAB International, Wallingford.

51. Ritchie, J.R.B., Crouch, G.I., 2000. The competitive destination: a sustainable perspective. Tour. Manag. 21 (1), 1–7.

52. RTP, 2002. Cape Town Declaration on Responsible Tourism. Responsible Tourism Partnership. Retrieved 15.6.2013 from: http://www. responsibletourismpartnership.org/RTWorld.html.

53. Sigala, M., 2008. A supply chain management approach for investigating the role of tour operators on sustainable tourism: the case of TUI. J. Clean. Prod. 16 (15), 1589–1599.

54. Stanford, D., 2008. Exceptional visitors: dimensions of tourist responsibility in the context of New Zealand. J. Sustain. Tour. 16 (3), 258–275.

55. STO, 2013. Green Tourism e Sustainable Development Framework in Slovenia. Retrieved 16.3.2013 from: http://www.slovenia.info/?ps_trajnost_ slovenija¼0&lng¼2.

56. Swarbrooke, J., 1999. Sustainable Tourism Management. CABI, Wallingford.

57. TSG, 2007. Action for More Sustainable European Tourism. European Commission, Brussels. Retrieved 5.12.2013 from. http://www.tolearn.uni-kiel.de/tolearn/ index.php?id¼128.

58. TSG, 2012. Charter for Sustainable and Responsible Tourism. European Commission, Brussels.

59. UNEP, WTO, 2005. Making Tourism More Sustainable. Paris: United Nations Environment Programme. World Tourism Organization, Madrid.

60. UNWTO, 2004. Indicators of Sustainable Development for Tourism Destinations. A Guidebook. United Nations World Tourism Organisation, Madrid.

61. Van de Ven, A.H., 2007. Engaged Scholarship: a Guide for Organizational and Social Research. Oxford University Press, Oxford.

62. van Marrewijk, M., 2003. Concepts and definitions of CSR and corporate sustainability: between agency and communion. J. Bus. Ethics 44 (2–3), 95–105.

63. Watts, R.E., 2011. Developing a conceptual article for publication in Counseling Journals. J. Couns. Dev. 89 (3), 308–312.

64. WCED, 1987. Our Common Future. Oxford University Press, Oxford.

65. Wheeller, B., 1993. Sustaining the ego. J. Sustain. Tour. 1 (2), 121–129.

66. Young, G., 1973. Tourism Blessing or Blight? Penguin Books, Inc., Baltimore.

67. Zuber-Skerritt, O., 1996. New Directions in Action Research Taylor & Francis. Retrieved 12.4.2014 from: http://proxy. library.brocku. ca/login?url¼http://lib. myilibrary.com?id¼2055.

2

CHAPTER

An Integrated Approach to "Sustainable Community-Based Tourism"

Tek B. Dangi * and Tazim Jamal

Department of Recreation, Park & Tourism Sciences, Texas A&M University, College Station, TX 77843, USA

ABSTRACT

Two rich knowledge domains have been evolving along parallel pathways in tourism studies: sustainable tourism (ST) and community-based tourism (CBT). Within both lie diverse definitions, principles, criteria, critical success factors and benefits sought or outcomes desired, advocated by different stakeholders ranging from quasi-governmental and non-profit organizations to public-private sector and academic interests. This poses significant challenges to those interested in theory building, research and practice in the sustainable development and management of tourism. The paper builds on a previous article published in Sustainability by presenting an integrated framework based on a comprehensive, in-depth review and analysis of the tourism-related literature. The study reveals not just common ground and differences that might be anticipated, but also important sustainability dimensions that are lagging or require much greater attention, such as equity, justice, ethical and governance issues. A preliminary framework of "sustainable community-based tourism" (SCBT) is forwarded that attempts to bridge the disparate literature on ST and CBT. Critical directions forward are offered to progress research and sustainability-oriented practices towards more effective development and management of tourism in the 21st century.

Keywords: sustainable tourism; community-based tourism; principles; governance; justice; equity; ethics; tourism development; tourism management

1. INTRODUCTION

A growing literature on sustainable tourism and community-based tourism has emerged over the past three decades in the field of tourism studies. While the discourse of sustainable tourism (ST) is oriented towards long-term sustainability, the literature on community-based tourism (CBT) looks towards local-level responsibilities and practices of tourism development and management. Numerous and diverse interpretations, definitions and practices have been forwarded within each along with a host of criticisms (see for a summary, for example, [1]). Progress towards sustainability is currently a rocky road littered with multiple definitions, indicators, stakeholders and principles [2,3,4,5,6,7,8]. No serious attempt has been made to systematically examine what they mean in relation to each other and to the overall goal of sustainability in tourism development and management, even though sustainability and "community" well-being are integral to both ST and CBT and offer common grounds for reconciliation and synthesis of these two complex domains. This paper undertakes a critical analysis of the relationship between ST and CBT, the principles and concepts that guide these approaches, their origin in institutional, historical and spatio-temporal basis and similarities and differences between the two approaches. The in-depth literature review shows steady progress being made in managing environmental, social and economic impacts, but clear directions for good governance, justice and ethics tend to be lagging in both the ST and CBT discourses. It is argued here that an integrated approach is needed to advance research and management practice past the current arrays of confusions and continued isolation between these two critical areas of tourism research. Such an integrated approach could help to better address management issues related to decision making and control over the fair distribution, use and conservation of resources and to achieve the desired goals of sustainable, community-based tourism that are claimed by both ST and CBT.

The purpose of this paper is therefore to examine and offer directions forward for an integrated approach to "sustainable community-based tourism". A comprehensive analysis of the relevant literature in tourism studies reveals that the rich knowledge domains of sustainable tourism and community-based tourism have been evolving primarily along parallel pathways in tourism. The paper traces some key developments and offers a framework to bridge these disparate discourses and to ensure that sustainability in tourism development and management is indeed grounded in community as a key principle (see the section on community-based tourism below for the meaning of "community"). The following research questions guide the study undertaken here:

(1) What is the relationship between ST and CBT?

- What principles and criteria guide these concepts as discussed in the tourism literature?
- How did they arise (institutionally, historically, spatio-temporally) and why?
- What are the similarities and differences between the two approaches?

(2) Is there a need to reconcile these two approaches or are they mutually exclusive?

- If integration is merited, what are the key principles or criteria that ought to guide such a reconciliation?
- Following from the above, what would an integrated framework of sustainable community-based tourism (SCBT) look like?

The paper is organized as follows. The Approach and Method Section below describes how the systematic examination of the research literature in ST and CBT was undertaken. An overview of the historic evolution of these two areas is presented in the next two sections. While some of this material is present in some form in the tourism literature, the context summarized here is important in order to understand the critique presented later in the paper. The subsequent section takes a closer look at some of the similarities and differences between ST and CBT and identifies a number of critical issues that are lagging in attention, yet are crucial in terms of sustainable development and management of tourism-related resources. A preliminary framework of "sustainable community-based tourism" (SCBT) is proposed. Critical directions forward are then offered to address a number of compelling issues that continue to hinder research and practice.

2. APPROACH AND METHOD

A comprehensive literature review and analysis of ST and CBT was undertaken to get a sense of the chronology (evolution) of these terms, key organizations and stakeholders involved, key criteria and principles used in certification schemes of ST and CBT, and tourism types and approaches (e.g., pro-poor tourism, responsible tourism, ecotourism). Firstly, some steps suggested for a scoping review [9] were followed, and a few relevant scholarly journals, book chapters, book reviews and conference papers were reviewed to get a closer view of the field and to identify the search terms. The search terms included: sustainable tourism; community-based tourism; responsible tourism; (sustainable tourism) (community based tourism) framework/model/criteria/indicators/principles/definitions/certifications. Given our own management background, the initial literature search in June 2014 used the commercial literature database Business Source Complete. This search produced 375 literature records. After duplicates were removed (due to the same records being retrieved through different keyword searches), 341 records remained for further screening. Of these records, only refereed journal articles and book chapters published in English were retained (content published in a language other than English, book reviews and conference papers were removed). Again, some of the methods of search, appraisal, synthesis and analysis (SALSA) recommended as part of the scoping review process [10] were applied to the 178 journal articles and book chapters retained. Further records were gathered based on insights and recommended readings that emerged during the review process (e.g., critiques of governance and justice, recommended publications from leading international institutions in tourism, such as United Nations World Tourism Organization and United Nations Environment Program). A few conference/seminar papers were included at this stage that had been repeatedly cited (for example, [11]). All abstracts were reviewed, and articles directly pertinent to the research questions were reviewed in depth to identify major themes, concepts and issues. In total, over 200 articles and book chapters were drawn up from this preliminary search that provided valuable material for the study and helped to inform the Scopus search below.

An expanded database search was subsequently conducted in the Scopus database in spring 2016, using a similar procedure as above. However, we added a new search category here, seeking to further ascertain the relationships between ST and CBT, as well as to further explore dimensions of governance, justice and ethics. A search using the words "CBT and principles" in the title, abstract or keywords resulted in 103 records of all document. A similar search including the words "responsible tourism and principles" resulted in 50 records, and "ST and principles" revealed 417 records. Limiting results to English language publications and book chapters or articles resulted in 257 records altogether. Finally, a focused search of the words (governance OR justice OR ethics OR equity) AND ("sustainable tourism" OR "community-based tourism"), following all the above methods, but unchecking one physical science subject area, came up with 170 articles and book chapters. Interestingly, 207 articles and book chapters from the new Scopus search were duplicated in our preliminary 2014 search.

3. RESULTS

3.1. Evolution of Sustainable Development and Sustainable Tourism

A number of global institutional initiatives based in Europe began to arise in the 1970s that shaped subsequent sustainability directions. The first United Nations Conference on the Human Environment was held in Stockholm, Sweden, in 1972. It established a place on the global agenda for environmental issues, for a "point has been reached in history when we must shape our actions throughout the world with a more prudent care for their environmental consequences" [12] (p. 1). In addition, the 1972 United Nations Educational, Scientific and Cultural Organization (UNESCO) Convention established guidelines for the protection of global natural and cultural heritage and required states to participate in the protection and conservation of officially designated World Heritage Sites [13]. Important contributors to the subsequent discussions and initiatives were the Club of Rome's report *The Limits to Growth* (for details, see [14]), and the *World Conservation Strategy*, 1980, was developed jointly by International Union for Conservation of Nature, United Nations Environmental Program and World Wildlife Fund (see also [15]).

The notion of sustainable development (SD) was forwarded by the United Nations World Commission on Environment and Development (UNWCED) in the seminal Brundtland Commission's report *Our Common Future* [16]. Sustainable development is "development that meets the needs of the present without compromising the ability of future generations to meet their own needs" [16] (p. 8). The report raised global awareness among public and private sector institutions for the need to consider a long-term conservation horizon, as well as societal considerations, such as: (i) intra-generational and inter-generational equity in the use and conservation of environmental resources; and (ii) north-south equity, *i.e.*, bridging the development disparities between the developed (Western) world and the lesser developed and poorer regions. This important initiative provided further momentum to growing concerns about long-term resource conservation and use.

Similar to sustainable development, sustainable tourism development, defined as a sub-set of sustainable development, witnessed some joint global institutional initiatives to direct it towards a balanced path even prior to the UNWECD (1987) initiatives. In the 1970s, UNESCO and the World Bank made an alliance for tourism development, the former supporting heritage preservation with expertise and the latter financing tourism-related infrastructure development. In 1976, these organizations jointly convened a seminar "to discuss the social and cultural impacts of tourism on developing countries and to suggest ways to take account of these concerns in decision-making" [17] (p. ix). However, the importance of addressing tourism as an important player in sustainability was not well recognized in the early initiatives mentioned above. Hall, Gossling and Scott [14] note that tourism was hardly mentioned in the UNWECD (1987) [16] report. The notion of "sustainable tourism" only later became engrained in the policy statements and planning documents of the United Nations World Tourism Organization (UNWTO), World Travel and Tourism Council (WTTC).

The role of tourism did arise at the UN Conference on Environment and Development (UNCED) held in 1992 (also known as the Rio Summit), which sought to help operationalize sustainable development through concrete, but non-binding actions (see [18]). The Agenda 21 action plan that resulted from the Rio Summit was adopted by 182 governments present at the UNCED conference. Tourism was identified here as "one of the five main industries in need of achieving sustainable development" [5] (p. 11). The World Travel and Tourism Council (WTTC), the United Nations World Tourism Organization (UNWTO) and the Earth Council (EC) subsequently jointly developed *Agenda 21 for the Travel and Tourism Industry: Towards Environmentally Sustainable Development* [19]. While Agenda 21 acknowledged only the potential of nature-based and low-impact tourism (ecotourism) enterprise, Agenda 21 for the Travel and Tourism Industry emphasized the need to make all travel and tourism businesses sustainable and detailed priority areas and objectives for governments and the tourism industry to comport with

Agenda 21. It called for travel and trade businesses in tourism to minimize negative impacts and forge partnerships for sustainable development, including collaborating with local communities. By this time, UNWTO had also developed a clear statement on sustainable tourism:

"Sustainable tourism development meets the needs of present tourists and host regions while protecting and enhancing opportunity for the future. It is envisaged as leading to management of all resources in such a way that economic, social, and aesthetic needs can be fulfilled while maintaining cultural integrity, essential ecological processes, biological diversity, and life" support systems [20] (p. 30).

As the link between poverty and environmental degradation became clearer, the United Nations World Summit on Sustainable Development held in Johannesburg in 2002 focused on poverty alleviation as a key priority. Again, the role of tourism in advancing social sustainability made significant headway under discussions of responsible tourism and pro-poor tourism. "*Making Tourism More Sustainable: A Guide for Policy Makers*" by UNEP-UNWTO [21] was a comprehensive policy document that described 12 aims of sustainable tourism development related to three "pillars" of sustainability: economic, social and environmental; also, see Figure 1 [21] (p. 20):

(1) Economic sustainability, which means generating prosperity at different levels of society and addressing the cost effectiveness of all economic activity. Crucially, it is about the viability of enterprises and activities and their ability to be maintained in the long term.

(2) Social sustainability, which means respecting human rights and equal opportunities for all in society. It requires an equitable distribution of benefits, with a focus on alleviating poverty. There is an emphasis on local communities, maintaining and strengthening their life support systems, recognizing and respecting different cultures and avoiding any form of exploitation.

(3) Environmental sustainability, which means conserving and managing resources, especially those that are not renewable or are precious in terms of life support. It requires action to minimize pollution of air, land and water and to conserve biological diversity and natural heritage [21] (p. 9).

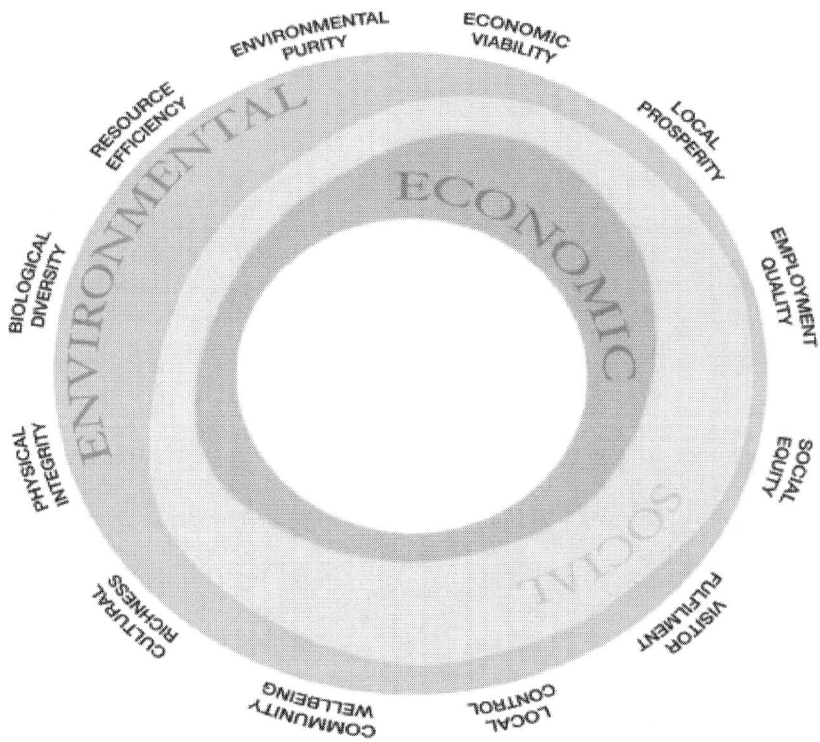

Figure 1. Aims and pillars (dimensions) of sustainability. Source: [21] (p. 20). © "United Nations World Tourism Organization (UNWTO), 92844/07/16".

While the earlier 1994 UNWTO definition [20] of ST mentioned "host regions", the 2005 report [21] more specifically incorporated "host communities", equity and cultural recognition (see the second pillar in Figure 1). It described sustainable tourism as "tourism that takes full account of its current and future economic, social and environmental impacts, addressing the needs of visitors, the industry, the environment and host communities" [21] (p. 12). ST was envisaged here as a continuous improvement process to be applied to all forms of tourism and all types of destinations and by the key stakeholders involved in ST: tourism enterprises, local communities, environmentalists, tourists and government. Their interests have to comport with the major aims of sustainable tourism, which were further specified in the chapter *Guiding Principles and Approaches of Sustainable Tourism Development* from the abovementioned policy document *Making Tourism More Sustainable: A*

Guide for Policy Makers [21]. Table 1 provides a chronological view of the evolution of SD and ST development through the institutional activities summarized above.

Table 1. Chronological evolution of sustainable development (SD) and sustainable tourism (ST) development.

Year	Evolution Points	Outcome	Citation
1972	UN Conference on the Human Environment in Stockholm, Sweden	Promoted the concept of eco-development integrating cultural, social and ecological goals with development	[12]
1972	Publication of the Club of Rome's report *The Limits to Growth*	Provides warning sign for economic growth-focused development	[14]
1972	UNESCO Convention Concerning the Protection of the World Cultural and Natural Heritage	Defined natural/cultural heritage and made state parties responsible for their protection and conservation	[13]
1980	Publication of *World Conservation Strategy*	Links conservation and sustainable development to human well-being	[15]
1980s	Rise of alternative approaches to and forms of responsible tourism in response to concerns about mass tourism	Ecotourism, responsible tourism, community-based tourism, pro-poor tourism, *etc.*	
1983	World Commission on Environment and Development (UNWCED)	Urged corporations and countries to engage in environmental conservation, pursue SD.	[16]
1987	Publication of Brundtland Commission report "Our Common Future" (UNWCED)	Coined and defined the meaning of the term SD	[16]
Early 1990s	Other definitions of ST emerge, broadly defined and reflecting alternative tourism efforts emerging in the 1980s onwards	"All forms of tourism which respect the host natural, built, and cultural environments and the interests of all parties concerned"	[11] (p. 480)
1992	UN Conference on Environment and Development (Earth Summit, Rio Summit	Rio Declaration on Environment and Development, Agenda 21, Principles towards implementing SD	[18]
1994	United Nations World Tourism Organization defines Sustainable Tourism	Emergence of ST at the global quasi-institutional level (emergence of global governance initiatives)	[20]
1995	Agenda 21 for the Travel and Tourism Industry	Urges governments and tourism industry to engage in sustainable tourism practices	[19]

2000	UN adopts eight (8) Millennium Development Goals (MDGs).	MDGs emphasize eradication of poverty and hunger, promotion of gender equality, and environmental sustainability among others; MDGs guide sustainable tourism development in UN member-nations	[22]
2000–2005	Tourism and climate change; slow tourism (2005) starts to show prominence	Tourism and global environmental change become hot-topics in tourism; detailed in the Routledge Handbook of Tourism & Sustainability	[23]
2002	World Summit on Sustainable Development (WSSD) or Rio+ 10, Johannesburg	Identified some deficiencies in the implementation of Agenda 21; delivered a political declaration, Johannesburg Plan of Action, and the establishment of partnership initiatives	[24]
2005	UNEP-UNWTO	UNEP-UNWTO form partnership to guide sustainable tourism development through recommending policies and tools	[21]
2010	Green growth and steady-state tourism enter sustainable tourism debate	Emphasis on sustainable consumption; detailed in the Routledge Handbook of Tourism & Sustainability	[23]
2012	United Nations Conference on Sustainable Development (UNCSD) or Rio+ 20, Rio de Janeiro.	Two themes were the focus: (1) a green economy in the context of SD poverty eradication; and (2) the institutional framework for SD; reaffirmed Rio principles and past action plans, and sustainable tourism was defined as one of the action areas	[24]
2015	UN Sustainable Development Summit, 2015; sets 17 Sustainable Development Goals (SDGs)/Global Goals	SDGs' target ending poverty, fighting inequality and injustice, tackling climate change, and so on, by 2030, building on the MDGs	[25]

A number of other initiatives are also worth noting. The UNWTO's Global Code of Ethics for Tourism (GCET) was adopted by the UN General Assembly in 1999. Its ten guiding principles were intended to facilitate tourism players in maximizing the positive impacts and minimizing the negative impacts of tourism on the natural and cultural environment in the world [26]. Articles 4 and 5 of the GCET highlight the benefits of tourism to host countries and communities and lay some obligations on its stakeholders. Yet another global system emerged in August 2010 as the result of a merger between the Global Sustainable Tourism Council (GSTC) and the Sustainable Tourism Stewardship Council (STSC) in

September 2009. This new GSTC umbrella organization has a diverse, global membership base, including UN agencies, leading travel companies, hotels, country tourism boards and tour operators, to foster sustainable tourism practices. Its criteria are intended to especially apply to hotels, tour operators and destinations and are organized under four pillars ([27], [28] (p. 217)):

(1) Demonstrate sustainable destination management

(2) Maximize economic benefits to the host community and minimize negative impacts

(3) Maximize benefits to communities, visitors and culture; minimize negative impacts

(4) Maximize benefits to the environment and minimize negative impacts.

The GSTC states that the destination criteria and performance indicators "were developed based on already recognized criteria and approaches, including, for example, the UNWTO destination level indicators, GSTC Criteria for Hotels and Tour Operators, and other widely accepted principles and guidelines, certification criteria and indicators" [27] (Section 2, Paragraph 4). While the GSTC's pillars and criteria are exemplary in addressing social and cultural impacts and issues, in addition to the interrelated environmental and economic interests that drove earlier initiatives, such as the UNWCED's Brundtland Commission report, critics point out that most of the certifications and initiatives forwarded tend to require voluntary compliance (aided by incentives) and continue to lack regulatory oversight and enforcement [29]. As Butler [4] and Wall [30] felt, sustainable tourism seemed to be more oriented towards sustaining tourism in the long-term (without adequate consideration of the trade-offs and costs involved).

3.2. Civil Society Organizations and Academic Sustainability

A number of civil society, non-profit organizations, as well as academic influences have emerged over time and much more quickly recognized the need to integrate social and environmental sustainability. The Brundtland Commission report [16], Agenda 21, the UNWTO (1994) definition [20] and the associated *Agenda 21 for Travel and Tourism* had focused primarily on environmental conservation in the early phases. By contrast, the fact that tourism drew on environmental and social-cultural good found in the public sector (not just under private property) raised tremendous challenges in terms of the public good and societal well-being.

Sexual and worker exploitation and mitigating the adverse impacts of sex tourism in places such as Bangkok, Thailand, were early concerns of the Ecumenical Coalition on Tourism in the 1980s [31]. Alternative approaches to tourism favoring small-scale, environmentally-friendly, locally-based tourism were advocated and seen to offer hope for locally-driven action and control over economic development, environmental and cultural conservation, as well as poverty alleviation and capacity building in the lesser developed regions in the south [32,33,34]. The notion of "responsible tourism" (RT) that arose in the 1980s fit the spirit of the alternative tourism movement of the 1970s [35]. The RT agenda was oriented towards local well-being, and according to Goodwin, "recognizes the importance of cultural integrity, ethics, equity, solidarity, and mutual respect placing quality of life at its core" [6] (p. 16). Emerging tourism forms included "soft and educational tourism (Krippendorf, 1982), cooperative tourism (Farrell, 1986), appropriate tourism (Richter, 1987), responsible tourism (Wheeler, 1991), special-interest tourism (Hall and Weiler, 1992), ecotourism (Boo, 1990; Ceballos-Lascurain, 1991) and... pro-poor tourism" as mentioned by Miller and Twining-Ward [36] (p. 31). Pro-poor tourism (PPT) arose as a specialized, small-scale, community-based approach to poverty alleviation in South Africa, spreading regionally, e.g., to The Gambia, Tanzania, Egypt [37,38,39,40] and further [41,42,43]. More recent forms such as agritourism (farm tourism), cultural tourism and volunteer tourism are evident in both the north and the south [29].

Following the global advocacy of sustainable development by the Brundtland Commission's report, the World Wildlife Fund commissioned Tourism Concern, a U.K.-based nonprofit organization, to prepare a discussion paper on "sustainable tourism" in 1992. Tourism Concern's report focused on both environmental and social sustainability, as the four key principles for sustainable tourism it forwarded shows: sustainable use of resources; reducing over-consumption and waste; maintaining diversity; and supporting local communities. It specifically recommended six processes: "integrating tourism into planning; involving local communities; consulting stakeholders and the public; training staff; marketing tourism responsibly; and undertaking research" [6] (pp. 14–15). The academic community followed suit. The launch of the *Journal of Sustainable Tourism* (JOST) in 1993 reflected rising interest in the academic community to undertake research and inform theory development and the practice of sustainable tourism. Its founding editors, Bill Bramwell and Bernard Lane, identified four sustainability principles in their first JOST editorial that similarly recognized the importance of addressing social well-being and identified four key stakeholders: the tourism industry, host communities, government and visitors [44].

Numerous research articles, critiques and reviews have followed the rise of the sustainable tourism agenda. A few noted early on the need to reconcile sustainable development and sustainable tourism development (see, for example, [45,46,47]. Farrell and Twining-Ward [48] presented sustainable tourism as a complex and dynamic system requiring careful management, including resiliency planning, risk assessment, hazard mitigation and adaptive planning. Collaborative planning for sustainable tourism and multi-stakeholder involvement are key principles for good governance and management [49,50]. Handbooks, such as the *Routledge Handbook of Tourism and Sustainability* (Edited by C. Michael Hall, Stefan Gossling and Daniel Scott) [23] and the *Routledge International Handbook of Sustainable Development* (Edited by Michael Redclift and Delyse Springett) [51], offer useful summaries and identify new areas of emerging study on sustainable tourism development and management, such as climate change and climate justice.

3.3. Community-Based Tourism: Evolution and Intersections

Weaver [52] (p. 206) observed that community-based tourism (CBT) was referred to in the early 1980s as the *sine qua non* of alternative tourism. Hopes were especially high of combatting mass tourism in the developing world and aiding rural communities in the global south through grassroots development, resident participation, empowerment and capacity building (see Scheyvens [41,42]). Peter Murphy's seminal book on community-based tourism [53], built on his research in the 1980s tourism in small communities in British Columbia and the Yukon (Canada). He advocated a systems perspective of community tourism where "community" very broadly refers to a group of people living in a defined space, and visitors interact with local people and landscape for a tourism experience ([50] (p. 188), [53])). We adopt his use of "community" for the purpose of this paper. Table 2 provides a few examples illustrative of the vast plethora of definitions and descriptions of CBT.

Table 2. Community-based tourism (CBT): some descriptions and definitions.

CBT	Author/Institute
CBT applies the objectives of ST "combined with an emphasis on community engagement and development".	[54] (p. 129)
CBT is an approach that engages the host community in the planning and development of the tourism industry.	[55,56]
"CBT is generally small scale and involves interactions between visitor and host community, particularly suited to rural and regional areas. CBT is commonly understood to be managed and owned by the community, for the community".	[57]
"CBT is a type of sustainable tourism that promotes pro-poor strategies in a community setting. The CBT initiatives aim to involve local residents in the running and management of small tourism projects as a means of alleviating poverty and providing an alternative income source for community members".	[58] (p. 10)
CBT refers to tourism that involves community participation and aims to generate benefits for local communities in the developing world by allowing tourists to visit these communities and learn about their culture and the local environment.	[43] (p. 2)
Community-based enterprises (CBEs) can be defined as a "Sustainable, community-owned and community-based tourism initiative that enhances conservation and in which the local community is fully involved throughout its development and management and they are the main beneficiaries through community development".	[59] (p. 637)

Despite the wide-ranging and widely different, often divergent political and cultural spaces in which CBT practices occur, common ground can be found with respect to the objectives and intended benefits of CBT, such as community development, capacity building, local control and local enterprise development, sustainable livelihoods and poverty alleviation (see [55,58,60]). CBT concerns and cares remain highly local, and the long-term sustainability vision articulated by the global institutions of the north is neither the goal nor the driver of "sustainable tourism" at this local level: community development, community survival, community involvement and local benefit are among the foci here. Vajirakachorn [61] identified 10 criteria for CBT success in her study of rural communities in Thailand: community participation, benefit-sharing, tourism resources conservation, partnership and support from within and outside the community, local ownership, management and leadership, communication and interaction among stakeholders, quality of life, scale of tourism development and tourist satisfaction. Based on a more regional study that included a wide range of case studies of village tourism, community tourism and ecotourism practices in various countries in the Asia-Pacific, Hatton [62] concluded that while the implementation and outcomes of community-based tourism vary, common themes are present, such as economic gain, leadership, empowerment and employment (see also [63,64]. The main foci were economic and social well-being.

Environmental sustainability and responsible use and management of environmental goods are not generally seen as dominant priorities unless conservation is a key issue, as in the case of ecotourism and community-based natural resource management (see [65,66]). Fennell [33] in a content analysis of 85 definitions of ecotourism found seven most frequently- and commonly-used elements including: nature based, conservation, reference to culture, benefits to locals, education, sustainability and impact. Pro-poor tourism (PPT) is an important form of CBT that emphasizes poverty alleviation with a specific focus on increasing the net benefits of tourism to the poor. Its goals include raising income thresholds and quality of life for poor and marginalized groups through capacity building and creating sustainable and diversified livelihoods (see [56], [58], p. 9). More recently, fair trade tourism has added greatly to the local and social focus of CBT, bringing issues of justice and fairness to workers in addition to the natural environment. Strambach and Surmeier [67] present a CBT-type program with "Fair Trade in Tourism South Africa" (FTTSA), established in 2001, and believe it to be one of the first innovative service standards that emphasizes social sustainability. The objective of FTTSA is to facilitate "a fair, participatory and sustainable tourism industry in South Africa" [67] (p. 740). Fair tourism business is defined as one that complies with the principles of fair share, democracy, respect, transparency, reliability and sustainability ([67] (p. 740), [68]). Examples from the developed world offer similar perspectives [69,70,71].

Closely related to achieving the community priorities noted above is the crucial area of "community"-oriented governance. Empowerment and resident participation are considered essential [69,72,73], and a key principle of CBT is that development and use of the community's goods and resources should be locally controlled, community-based and community driven. Yet, within the vast literature on CBT, while community involvement and resident participation are relatively ubiquitous principles, community *ownership* and resident *control* over decision making face significant challenges, and examples of CBT success are sparse. In a critical review drawing from various "community-based and -driven" development projects from around the world, Mansuri and Rao [74] offer valuable insight into the crucial political (and cultural) issue by carefully distinguishing *community-driven* development from *community-based* development:

Community-based development is an umbrella term for projects that actively include beneficiaries in their design and management, and community-driven development refers to community-based development projects in which communities have direct control over key project-decisions, including management of investment funds.

[74] (pp. 1–2)

Scheyvens [41,42] identifies four dimensions of community empowerment: economic (income and employment related); psychological (considers community pride and self-esteem); social (community cohesion and well-being); and political (shift balance between the powerful and powerless, between the dominant and dependent, for greater political equity). Resident empowerment through tourism scales that were developed and tested in Virginia, USA, by Boley, Mcgehee, Perdue and Long [75] offers corroborative support for these economic, psychological, social and political empowerment dimensions. Table 3 summarizes a number of critical success factors (CSFs) for CBT that are understood to be common worldwide. These are organized under four key areas of community empowerment identified by Scheyvens [41,42] (see also [60,75]). The geographically-diverse scope of CBT spans the developing to the developed world (see, for example, [43,59,61,64,70]). Some scholars have identified additional criteria for inclusion that are generally not typical of CSFs, but important in terms of sustainability, tourism, as well as community well-being. These include environmental protection and management (e.g., community-based natural resource conservation) [66], infrastructure development [63,71] and direct control over the distribution of CBT benefits [74].

It should be noted that the CSFs listed in Table 3 are not fully comprehensive, but are representative of commonly cited CSFs in the tourism development and management literature. Numerous other criteria have been proposed less frequently for inclusion, such as environmental protection and management, infrastructure development [63,71], distribution of equitable benefits [74] and flagship or other significant (mega) attractions to support rural tourism development as a growth pole [70].

Table 3. Some commonly-cited critical success factors (CSFs) organized under community empowerment dimensions provided by Scheyvens (1992/2002).

Dimensions of Community Empowerment	Elements of Community Success Factors (CSFs)	Key Authors/Sources
Economic Empowerment	**1. Income and employment** 1.1. Economic benefits 1.2. Local ownership of businesses, small and medium business enterprises (SMEs) 1.3. Providing financial services/funds to SMEs 1.4. Management of external/internal financial resources 1.5. Vision, goals, strategies, marketing/networking (integrated planning) 1.6. Economic benefits 1.7. Capacity building, training and entrepreneurship/skills development 1.8. Equal distribution of land among residents/equity 1.9. Community assets	[41–43,59–61,64,70,71,76–78]
Psychological Empowerment	**2. Community pride and self-esteem** 2.1. Participation, involvement, collaboration 2.2. Educational activities (to identify self needs), having knowledge/information 2.3. Tourist/resident satisfaction 2.4. Protecting local identity	[41,42,50,60,61,66,75,76,78,79]
Social Empowerment	**3. Community cohesion** 3.1. Participation, involvement, collaboration 3.2. Community cohesion, networking, sense of community 3.3. Interaction among stakeholders 3.4. Quality of life 3.5. Respect for local culture and tradition 3.6. Tourism resource conservation 3.7. Important role of women in development	[41–43,56,59–61,63,69,72,73,75–78,80]
Political Empowerment	**4. Shift in power balance** 4.1. Participation, involvement, collaboration 4.2. Support from local/national government 4.3. Visionary and passionate leaders	[41,42,59–61,63,64,75,80]

3.4. Comparing CBT to ST: Similarities, Differences and New Insights

The role of the local communities, while ambiguous or little emphasized in the institutional beginnings of SD and ST, became quickly evident in later initiatives like *Local Agenda 21* and in various UNWTO documents (Figure 1 [21]). As Table 2 above shows, direct links have been made between ST and CBT. Some argue that ST principles are applicable to all forms of tourism, whether mass tourism, alternative tourism or CBT [21,46]. However, a careful comparison between the two raises important considerations for the development and management of tourism, ranging from conceptual and theoretical issues to issues of scale and size, as well as the engagement of the public/private sectors and the role of residents in matters of public good and societal well-being.

The literature review reveals a proliferation of stakeholders and interests in ST and CBT and a related abundance of definitions, criteria and principles related to sustainability. This makes good planning, implementation, management and monitoring problematic in the complex tourism system [48], both in research and practice. *Conceptual and definitional challenges* abound. Graci and Dodds [5] claim that there are over 200 definitions of sustainable tourism, without an internationally-recognized one. Weaver [7], for instance, championed sustainable mass tourism (SMT) to meet the visitor/destination demand and economic growth to be operationalized by applying three paths: (1) organic (spontaneous, market-led growth); (2) induced (mix of regulated and market-led growth); and (3) incremental (regulated) (CF [81]). Additionally, as in the ST literature, numerous definitions and approaches proliferate in the CBT literature, and little consensus appears to exist on the key principles of sustainability. While numerous certification and accreditations schemes have arisen in the context of sustainable tourism, ecotourism, rural tourism, *etc.*, their effectiveness is inhibited by the primarily voluntary nature of these programs, lack of strong oversight and regulation and the global mobilities of transnational organizations, labor, capital, *etc.* [82]. Local resident control is seen to be a losing battle [83].

The scale and scope of tourism and the numerous stakeholders that drive ST and CBT make it extremely difficult to manage the local to global commons for environmental, social and cultural sustainability: *good governance* remains problematic at all levels from the local, to the global, and climate change and extreme weather will continue to exacerbate the challenges (see [84]). Initiated mainly by international public-private institutions (such as the United Nations' Earth Summit, UNWTO, UNEP, WTTC and various tourism scholars especially originating in West-Europe, North America, Australia and New Zealand), ST is generally conceived of on a larger scale than the local (historically regional/global). Its discourse is criticized as Eurocentric and overly focused on, as well as influenced by business interests [6,8]. By contrast, CBT has its origins in the local, focusing on grassroots development through participation, equity and empowerment and emphasizes local enterprises developed through local knowledge and entrepreneurship [43]. In the developing world, specifically the global "south" regions of Africa, Asia, South America, the origins of CBT can be traced to rural community development, economic and capacity building, social justice, poverty alleviation and community-based conservation, while economic considerations primarily drove CBT in the global "north" [6,8,29,85]. Similar to ST, CBT is also promoted by conservation, non-governmental, and international organizations, such as the World Bank and the Global Environmental Facility [86] (p. 2). However, it aims to maximize benefits for community stakeholders rather than absent investors, who may still be participants in ST operations, such as airlines, mega-resorts and chain-hotel investments. CBT initiatives are small enterprises. In this respect, CBT also overlaps with responsible tourism (RT) to seek an approach that "benefits local community, natural and business environment and itself" [87] (p. 314). CBT thus very clearly identifies with distributive and social justice, ethical relationships and equity, from its rootedness in the locale/community. CBT and ST thus reflect approaches with substantial intersections and overlaps, but also some significantly diverse foci.

However, as discussed further below, good governance is a significant challenge at the local level of CBT as much as at the larger macro-level ST discourses noted above. The political economy of tourism in the developing world, as Britton's well-known article demonstrated, is driven by powerful external stakeholder, market capitalism and neoliberal interests [88] (see also [89]). Table 4 compares and summarizes some of the issues and challenges that face these two crucial discourses of sustainability.

Table 4. ST and CBT: comparisons and contrasts.

ST	CBT
Macro-level, regional- to global-scale concerns over long-term environmental sustainability and development by quasi-governmental, global institutions, like the United Nations, with UNWTO, UNEP, World Travel and Tourism Council (WTTC) playing an active role in tourism. Other non-profit and membership-based institutions have arisen over time (see [27,84].	Non-profit and public sector influence at the local level; NGOs, local/national government, small practitioners/community, donor agencies and tourism scholars [6,77].
Eurocentric discourses and industry interests drive planning and policy making on resource use and conservation; societal and public good are a secondary concern, though equitable benefits for current and future generations are stressed. ST implementation effectiveness is seen to be poor [6,8,29,85,88].	Grass-roots, subsistence/rural economies and local considerations drive CBT in developing regions (e.g., Asia, Africa, Latin America) and in Europe, America, *etc.* Different forms arise based on the main concerns and goals, e.g., pro-poor tourism (poverty alleviation), ecotourism, sustainable livelihoods and community-based conservation [43,59,66].
Industry actors are related to market capitalism and larger tourism and hospitality interests, such as airlines, hotels, large tour operators, *etc.*; critics argue that the primary concern is sustaining tourism [4,6,29].	Industry here is micro-level, practical, emphasis on local enterprises and local economic development, social well-being, cultural survival, poverty alleviation, capacity building [41,42]; community-driven and community-based economic and social priorities, grassroots development, resident participation, equity, empowerment, local entrepreneurship and small-scale enterprise [43,59].
Proliferation of definitions, principles, criteria forwarded by academic and other ST advocates; communication gaps between ST research in academia and ST industry efforts [85,89,90].	CBT definitions also proliferate; principles of CBT vary depending on source and interests. Synergies between CBT, PPT and ecotourism are discussed, but common principles are not clearly identified.
Critique by academics that ST certification is primarily voluntary and there is little regulation or oversight of the industry by government [29]. Monitoring and evaluation of ST use primarily quantitative indicators, measures and positivistic approaches, with little consideration for intangible aspects, such as cultural change [1].	Qualitative indicators plus quantitative to monitor and evaluate social sustainability, social and environmental justice, improving well-being of marginalized, disadvantaged groups, racism, gender equity, democratic participation and local control, social cohesion and inclusion of local knowledge [91,92].

3.5. A Closer Look at Some Key Dimensions of Sustainability

The review and analysis above reveal an important (and urgent) need to better integrate ST and CBT, for there are many similarities despite the difference of scale and priorities. Yet, even the priorities are not so dissimilar upon further examination, if one believes that individual and social well-being are integral to sustainability. The micro-level discourses of CBT help to "operationalize" some of the grander visions and goals of macro-level sustainability. Jamal, Camargo and Wilson [1] indicate the need for this as well. However, the vital role of governance at the community level is of crucial importance for successful CBT. How does this relate in the context of ST? Keeping this in mind, further examination of the literature gathered helped to identify a number of under-represented issues that ought to be vital to the success of ST development, such as governance-related issues that are clearly evident in CBT. Table 5 shows a comparison of key dimensions identified in ST. The last row in Table 5 summarized a number of governance-related concepts that have been discussed more broadly in the tourism literature without using the label of ST (hence, these did not show up in our literature review), but which are of relevance to considerations of good management and governance.

Table 5. Some key ST dimensions and items that merit further consideration.

Pillars/Dimensions of Sustainability	Description	Organization/Scholar
Three (3) pillars or dimensions	Economic, social and environmental sustainability	[21]
Three Policy Issue Areas	Environmental, economic and social	[93]
Triple Bottom-line	Social, economic, environmental dimensions	[94]
Three dimensions	Physical/environment, socio-cultural, economic	[95]
Four (4) dimensions	(1) Sustainable destination management (2) Maximizing economic benefits to the host community and minimize negative impacts (3) Maximize benefits to communities, visitors and culture; minimize negative impacts (4) Maximize benefits to the environment and minimize negative impacts.	[27,28]
Four (4) dimensions	Environmental, economic, socio-cultural and institutional/management	[96,97]
Four (4) dimensions/contexts	Social, economic, environmental, governance/political	[98,99]
Four (4) dimensions	Environmental integrity, economic prosperity, social equity, tourism value chain	[100]
Four dimensions (4)	Social, political/administrative, physical environment, local environment	[101]
Issues not well addressed in the four (4) dimensions	Issues of governance, such as accountability, transparency, participatory governance (direct participation, empowerment to enable voice in decision making, planning and development of tourism) ethical principles related to justice and equity.	[1,3,10,82,83,92,98,102–109]

As the review conducted earlier in the paper shows, the literature is consistent with respect to identifying economic, social and environmental sustainability as key pillars/dimensions of ST (see [21,93,94,95]. Though the key dimensions remain the same, however, some specific details or extra dimensions are being added by other scholars and institutions over the years. For example, GSTC [27] adds one new dimension of sustainable destination management, which is missing in UNEP and UNWTO [21] and in the work of other scholars [93,94,95]. Moreover, all aspects of sustainability are not equally emphasized; some aspects have been mentioned by a few, omitted by many. Our examination of the literature also indicates that some important parameters are under-represented and poorly addressed in tourism research. For example, governance (institutional arrangements) has been mentioned by Bramwell [98], Roberts and Tribe [96] and Pukhakka, Cottrell and Siikamaki [97], with specific reference to sustainable tourism. García-Melón, Gómez-Navarro and Acuña-Dutra [101] mentioned government as a political/administrative dimension. Equity has been raised in the UNEP and UNWTO [21] aims and mentioned by Sharpley [47], Pomering, Noble and Johnson [100] and in FTTSA [68], but there is a paucity of good research on inequities in the distribution of goods, services and income and related issues of distributive and procedural justice. Jamal, Camargo and Wilson [1] also pointed out the need of a framework that emphasized justice and ethics as a new direction to sustainable tourism development. Jamal and Camargo [109] provided an introduction to destination justice that emphasized the need to address issues of equity and fairness. While they drew on Rawls' theory of justice, they argued that it is inadequate to guide a just destination, and further research was needed to better identify guiding principles for destination justice. Inter- and intra-generational equity and equity among the nations of the north and the south is one of the objectives of the Brundtland report [16]. Unfortunately, benefits to disadvantaged groups through destination justice have been grossly neglected in many works, excepting a few documents, such as *Guide for Local Authorities on Developing Sustainable Tourism* [110] (p. 43), FTTSA [68] and the work of Lee and Jamal [108], who emphasized the need for procedural and distributive justice to the justice to the disadvantaged. The Mohonk Agreement [111] exclusively mentioned ethical issues in business operations, but progress in theoretical and empirical research is slow, as a number of authors have noted (see the last row of Table 5).

In the case of sustainable tourism development, many researchers seem content with presenting inter-relations among commonly-understood dimensions, such as economic, socio-cultural and environmental [22,93,94,95], and proposing criteria under these three dimensions. However, there are some like [27,28,96,97,98,99,100,101,112] who find institutional mechanism/governance/political administrative aspects to be critical for the success of ST/CBT and other alternative forms of tourism, such as ecotourism, rural tourism, responsible tourism, *etc.* Siow May, Abidin, Nair, Ramachandran and Shuib also recommended institutional framework criteria under the socio-cultural dimension [95] (p. 149). The governance-oriented scholars acknowledge the importance of institutional mechanism, whether formal or informal, networked or hierarchical, in achieving desired sustainable and community development goals. For example, in the 7Es of sustainable tourism planning (environment, economics, enforcement, experience, engagement, enquiry, and education) in Catibag-Sinha and Wen [113], the important role of governance is also denoted: enforcement and engagement. Ellis and Sheridan [114] emphasized the active role of government in bridging the theoretical and practical challenges that might arise between external stakeholders (academics, NGOs outside of communities) and local stakeholders (NGOs and communities), especially in lesser developed countries. Drawing from a case-study of Kumarakom, India, Chettiparamb and Kokkranikal [87] argued that besides voluntary codes of conduct by/for the tourism industry itself, some sort of regulation, enforcement and coercion is needed from the government in providing social equity and community well-being for responsible/sustainable tourism to be effectively operational. Wight's [115] earlier study of the Province of Alberta, Canada, argued that an active government role was more effective in balancing tourism development goals harmonizing society, economy and environment rather than economic-focused goals led by the private-sector marketing body. Ottenbacher, Schwebler, Metzler and Harrington's [116] examination of sustainability criteria in Baden-Württemberg, Germany, similarly showed strong support for government to conduct/implement sustainability checks for highly diverse tourism attractions in the state. These and other similar studies support the central idea that any framework, criteria or model of ST/CBT development remains incomplete without the inclusion of governance and its role as a facilitator and regulator of tourism development and management through active engagement and responsible partnership with the private and non-profit sector as needed.

In addition to the above, the literature review we conducted also revealed that sustainable tourism lacks a uniform definition and draws from many perspectives in defining its concepts, dimensions and approaches. Likewise, the processes of criteria and indicator development of ST and CBT have remained elusive. Numerous criteria and indicators of sustainable tourism have been forwarded that are at times highly ambiguous or contradictory, and the overall result seems to offer more confusion than guidance. We offer below a preliminary framework of "sustainable community-based tourism" to integrate and offer direction forward into an approach that bridges the local and the global and that situates sustainable tourism within the context of the local (in terms of place and community).

4. DISCUSSION

4.1. Towards an Integrated Approach to ST and CBT

The review conducted thus far indicates that while some of the aims of ST and CBT appear to be similar (community and environmental health and well-being), there are significant differences and issues that are related to the spatial-temporal scale of this highly complex domain. Global issues of long-term planetary sustainability juxtaposed alongside local community/societal well-being require addressing diverse political, geographic and cultural contexts and stakeholders with varying interests and knowledge (public-private and non-profit, academic, residents, tourists, *etc.*). It is not surprising to see the plethora of definitions, principles and "success factors", but the answer is not simply to continue business as usual, *i.e.*, study and practice of ST and CBT as two different phenomena of long-term sustainability (ST) and local community well-being (CBT), as some might argue. Building on the previous section, this section teases out some of the more serious implications of the comprehensive literature review conducted above. A preliminary "synthesis" is then offered that may offer helpful directions forward to more effectively accomplishing mutually-desirable goals of the long-term sustainability of planetary resources and societal (including community) well-being.

4.1.1. The Global and Universal *versus* the Local and Particular

Following early drivers on sustainable development [16], a global vision and higher level of abstraction situated the emergence and development of ST, aiming towards common priorities of long-term planetary and resource sustainability along with other goals. Guidance (as most ST programs are voluntary) has generally been provided by global organizations and a number of smaller voluntary or non-profit organizations such as the NGOs, Netherlands Development Organization (SNV) and Tourism Concern. Global organizations include the UNWTO, WTTC, UNEP and GSTC, along with other institutionalized forms (such as national and regional governments, global certification programs, such as Green Globe, global advocacy organizations, such as Sustainable Travel

International). While the UNWTO attempts global governance through mechanisms, such as the Tourism Bill of Rights, the Global Code of Ethics and various related committees and initiatives, none of these have regulatory "teeth", and the tourism system continues to operate mainly on a self-regulatory system of voluntary efforts at responsible practices for sustainable tourism.

However, the challenges for sustainable tourism development do not end here. A variety of rights-based discourses are embedded in UNWTO documents, such as the Global Code of Ethics (see, for example, Article 6 on the right to tourism). Its own role is embedded in Article 3 in the Global Code of Ethics as developing and promoting tourism for economic development, understanding, peace, universal respect and observance of human rights and fundamental freedoms. This puts it in a powerful position as one of the global moral arbiters of "sustainable" tourism, which, of course, is embedded deeply in a discourse of rights and ethics. As Smith and Duffy [104] and others have pointed out, however, the universal rights discourses that emanate from dominant institutions like the United Nations (e.g., its global human rights declaration) and the UNWTO are historically and culturally embedded in modernity, imperialism and colonialism, situated in the global "north" (*i.e.*, "the West"). As various definitions of ST, CBT and the proliferation of goals, principles and criteria from the literature review show, the interests of Western, "developed" countries are oriented towards successful capital markets, profit and a use and development discourse embedded in instrumental use and material values of success and benefits (see Table 2, Table 3 and Table 4 for examples). Universal, rights-based discourses tend to favor such ends, as exemplified by justice theories, just as that of John Rawls [117], that emphasize distributive justice in Western liberal democracies.

By contrast, the interests of the developing world tend towards very particular political and social concerns, such as self-determination, autonomy from colonial and other oppressive political regimes, capacity building and socioeconomic development. Such diverse cultural, geographic and political values and interests challenge the universalist ethics that attempt to institute justice and governance across the board, at all levels and all scales from the global to the local. In between the local and the global lies a disparate set of stakeholders, including national, state, regional governments and other public/private sector and non-profit organizations, including multiple levels of destination management organizations (DMOs) with each country (national and regional DMOs, local convention and visitors bureaus, *etc.*). Feminist critiques and other critical perspectives argue that "difference" is ruled out by dominant, universalist rights-based discourses. Individual agency, as well as community control are matters that must be negotiated not just at the international level, but also at the local level, if respect, recognition, equity and justice are to be enabled for diverse groups, cultural views, *etc.* Feminist theories, as Smith and Duffy [104] note, call for an "ethic of care" that addresses difference and the particular (local, individual, for example), rather than universal, generalized theories of justice.

4.1.2. Under-Representation of Equity and Justice

Despite the clear call for intra- and inter-generational equity in the Brundtland Commission's global report on sustainable development [16], justice-related issues (e.g., distributive justice and procedural justice, rights of disadvantaged and diverse populations, minority ethnic groups in tourism destinations) have received limited attention in tourism studies [1] (p. 4605), but some momentum is beginning to gather (see [42,102,104,105,106,107,108,109,118,119,120,121]). Higgins-Desbiolles [106] argues justice tourism only as the true form of alternative tourism that has the capability of thwarting capitalists' interests as "it seeks to reform the inequities and damages of contemporary tourism...to chart a path to a more just global order" [106] (p. 345). Justice tourism is described as "both ethical and equitable", and it consists of attributes, such as that it: "(1) builds solidarity between visitors and those visited; (2) promotes mutual understanding, and relationship based on equality, sharing and respect; (3) supports self-sufficiency and self-determination of local communities; and (4) maximizes local economic, cultural and social benefits" [42] (p. 104). We draw from these authors to commence the discussion on equity and justice (Table 6), but much needs to be done in this knowledge domain. It is not surprising that Macbeth [105] proposed sustainability and ethics as the fifth and sixth platforms to supplement Jafari's existing four platforms in tourism studies (advocacy, cautionary, adaptancy and the knowledge-based platform).

4.1.3. Bridging the Global and the Local: Towards a Preliminary Framework

The issues and barriers identified above point to an urgent need for a more robust framework that can address both ST and CBT from an integrated approach: neither ST alone, nor CBT alone are sufficient. The global environmental need for planetary sustainability that has driven the discourse of ST has to be reconciled or brought into better synergies with the local priorities of community control, capacity building, empowerment and sustainable livelihoods. The global imperatives of climate change and poverty alleviation call for urgent new directions and innovations to meet the interdependent priorities of planetary sustainability and sustainable livelihoods. However, the scoping review shows an overwhelming number of conceptions, definitions and descriptions of ST and CBT, forwarded by a diverse number of stakeholders worldwide. The field is muddied with confusing terminology, a cornucopia of principles that do not appear to differentiate management principles from moral principles. Overall, the development of effective sustainability-oriented frameworks, policies and practices appears to have proceeded in separate academic and practice domains with respect to both ST and CBT.

Table 6. An integrated framework of sustainable community-based tourism (SCBT): dimensions and examples of criteria.

Dimensions	Examples of SCBT Criteria Drawn from the Literature Review	Authors/Sources
Economic	**Economic Benefits:** Capturing economic benefits; sustainability of tourism operations and services; economic monitoring; economic vitality; business performance/profitability; local economic development; economic well-being; local government income; rural development; national economic development; property values; local economic diversification; increased consumption of local products; supporting local entrepreneurs and fair trade; investments; employment; quality of employment; business motivation; revenue generation; business performance; income distribution/capital leakage and linkage; income and total sales; empowerment; local enterprise and ownership; sustainable livelihoods framework (SLF); local control	[27,43,58,66–68,86,96,122–129]
	Local Jobs and Participation: Local career opportunities/employment; public participation; local community opinion; local access; tourism awareness and education; support for community; labor/company and job conditions	[27,43,96,123,127]
	Institutional Mechanism to Ensure Economic Benefits: Fair wages; internal-external business operations; Income distribution; capital formation in the community/investment; nature of (visitors) demand; labor/company and job conditions; micro-credits; preventing exploitation; foreign exchange leakage and domestic linkages; demand and supply of local services; accommodation capacity; wages evaluation; tourism employment index; tourist expenditure pattern; index of foreign exchange revenue; integration of tourism with other activities	[27,43,58,66–68,93,96,122,123,125–129]
	Visitor Management: Seasonality; Length; Visitor expenditure	[27,43]

Table 6. *Cont.*

Dimensions	Examples of SCBT Criteria Drawn from the Literature Review	Authors/Sources
Environmental/ Ecological	**Protection of Natural Environment:** Protection of valuable natural assets; managing scarce natural resources (water availability and conservation; drinking water quality); protection of sensitive environments; wildlife protection	[27,68,96,122–126]
	Reducing Waste/Emissions: Limiting impacts of tourism activity (sewage treatment; solid waste management; greenhouse gas emissions; energy conservation; wastewater; solid waste reduction; light and noise pollution; recycling and reuse; pollution effects management; visual pollution (conformity to local vernacular); respect environment; rate of ecosystem destruction/degradation;	[27,58,68,96,122–124,126]
	Innovating/ Adaptive Planning to Environment-friendly Plans: Green design, permaculture gardens; alternative energy; vegetation; conservation zone; fostering human environment relationships; low-impact transportation; ecosystem; atmosphere; energy; resilience and risk; environmental awareness and management; energy efficiency	[27,96,123,124,126]
	Assessment and Monitoring: Environmental risks; assessment of environmental impacts of tourism activity; health of human population (residents/visitors); air; geology and soil; coastal and marine resources; environmental awareness; biodiversity and ecological health; natural capital; loss of renewable/non-renewable resources	[27,96,123–126]
Social-cultural	**Community Well-being and Satisfaction:** Community well-being (local satisfaction with tourism; effects of tourism on communities); sustaining cultural assets; attraction protection; crime and harassment; cultural promotion; ownership patterns; resident views (satisfaction); host reactions to tourists; local culture/tradition; community development; social cohesion; sex tourism; community resource; distribution of resource/power; community health and safety; quality of life in general; building/architecture; socio-cultural fabric; recreational quality; address conflicts of interests	[27,43,58,93,96,122–125,127]
	Community Participation and Empowerment: Community/resident involvement and participation in tourism; community assets/skills/involvement; uniqueness; community empowerment; cultural education; education and training; equitable changes in local lifestyle; site interpretation; intellectual property; distinction (visit to heritage sites); accessibility; resident access to tourism goods and benefits	[27,43,58,93,96,122–125,127]
	Visitor Satisfaction: Tourist satisfaction; visitor management; visitor behavior; accessibility and convenience	[27,122–124,128]

It is argued here that "sustainable" tourism development must be approached from an integrated macro-micro perspective, particularly in what appears to be increasingly globalized neoliberal development agendas [1,6]. As the above discussion indicates, issues of *justice*, *equity* and *fairness* in the distribution and use of tourism-related resources must be better addressed in the SCBT framework, both at the global and local level. Global governance and universal justice are needed, but so are local governance and attention to inequality, diversity and disadvantage at the local level. The UNWTO's Global Code of Ethics and various industry-related codes of ethics and ecotourism/sustainable tourism certification programs have arisen and are important to enable some accountability towards responsible behavior. Such codes remain primarily voluntary and self-regulatory, as noted earlier. Guidance from global organizations and global non-profit organizations, certification programs and associations can help the "self-regulating" tourism operators and industry stakeholders with universal, generalized principles and criteria. However, the local must participate in enabling good governance at the local level, facilitating environmental stewardship, social justice, well-being and sustainable local livelihoods. Culturally-appropriate mechanisms for resident involvement in tourism development, planning and decision making are needed that enable local control and good governance, ensuring justice, equity and fairness in the use and distribution of tourism goods. Principles of sustainable tourism in general include: (1) taking a holistic view; (2) pursuing multi-stakeholder engagement; (3) planning for the long-term; (4) addressing global and local impacts; (5) promoting human heritage and bio-diversity; (6) preserving essential ecological processes; (7) ensuring equity; and (8) supporting local communities [6,21,44]. Such principles are helpful to local level community-based enterprises and community development goals in CB. However, weaknesses in the implementation of these principles, whether owing to poor linkage or lack of effective governance and attention to issues related to justice, ethics and equity, remain a major obstacles. The integration of these principles into a framework of SCBT that supports the established environmental, social and cultural criteria with clear, strong emphasis on governance, justice and ethics (Table 5) is a crucial endeavor if major sustainability challenges (such as climate change that threaten further inequities and injustices to the poor and marginalized) are to be effectively addressed.

Table 6 and Table 7 together offer a preliminary framework for SCBT based on the scoping review. Key criteria and items common to ST and CBT that were obtained from the extensive literature review are listed as examples under the economic, environmental/ecological and social-cultural dimensions (see the second column; pertinent authors/sources are shown in the third column). All of the criteria and items in the second column of the framework are adopted from the authors/sources cited. However, to provide more legibility and uniformity to the framework, we have invented broad sub-headings (in bold) in the second column under which various criteria and items have been grouped.

Table 7. Governance and underrepresented issues in the domain.

Dimensions	Examples of SCBT Criteria Drawn from the Literature Review	Authors/Sources
Governance	**Planning/Strategic Vision:** Controlling tourist activities and levels (controlling use intensity); destination planning and control (development control); designing products and services; strategic plan; access to finance	[43,96,100,122,123,130,131]
	Management and Marketing: Sustainable destination strategy; destination management organization; tourism seasonality management; climate change adaptation; property acquisitions; promotion; promotion of ecotourism (eco-label); marketing (return-visits); tourist traffic (volume); management and staff training; partnerships (Government/Private Sector, NGO); local authorities encourage community participation; external advice and links	[27,43,86,97,100,122,123,126,132]
	Power, Rules and Regulations: Governing; steering; mobilizing; monitoring; inventory of tourism assets and attractions; planning regulations; access for all; visitor satisfaction; sustainability standards; monitoring and evaluation	[27,43,123,128]
	Visitor Safety and Crisis Management: Safety and security; crisis and emergency management; fair and safe working conditions; traffic	[27,68,123,126]
	Collaboration/Coordination: Sharing knowledge, thought, resources, power-sharing, multi-level integration, engagement, story-telling and pluralistic dialogue	[82,98,133]
	Participation: Involvement, diversity and decentralization; deliberation; transactional relationship; consensus orientation	[131,133]
	Service Delivery: Flexibility, revisibility	[131,132]
	Accountability: Responsiveness, efficiency, effectiveness	[82,132]
	Transparency: Clear operational structures and processes	[82,132]
	Equity: Pursuit of equity and inclusiveness; Acceptance of diversity	[82,132]
	Communication: Constructive information flow	[82,134,135]
	Leadership: Visionary	[132]
	Political: Local oriented control policy; political participation; local planning policy; political support at all level of governments	[124]
	Technological: Accurate data collection and tourism information change; adoption and use of new and low impact technologies; benchmarking-generic and competitive	[124]
	Underrepresented Issues in Governance: Equity, Fairness and Justice	
	Justice in Tourism: Inter-and intragenerational equity; equitable distribution of costs and benefits, goods and services; distributive justice benefiting disadvantaged populations; respect and recognition of diverse values; north-south equity, self-determination and autonomy of indigenous people; environmental and social-cultural justice; destination justice; address discrimination, racism, inclusiveness, human rights, *etc.*	[1,16,21,58,67,68,83,102–106,108, 109,117,118,126,135]
	Equity and fairness: Fair distribution of goods and resources; equal employment opportunity for all including women, youth, disabled and vulnerable population; poverty reduction; gender equity and social inclusion; fair wages and employment; respect and enable human rights; affordability and access (services targeted to low income, poor and disadvantaged populations).	[1,21,58,67,68,104,106,107]
	Related ethical issues: Understanding and applying moral/ethical principles in tourism Address intrinsic and instrumental values: utilitarian ethics; virtue ethics; respect for persons: Categorical Imperative (Kant), Ethics and the "Other"; feminist ethics, ethic of care, *etc.*	[1,104,108]

The literature reviewed above shows some ubiquitous goals and criteria that are common to both CBT and ST. Some governance-related items are consistently mentioned (e.g., local participation in decision making is a key principle of CBT and is noted by UNWTO for ST. However, the mantra of local control and resident-driven decision making in CBT tends to be problematic, as local governance is shaped and influenced by diverse historical, political, cultural and ethnic values. Capacity building, a key tenet of CBT, is also a challenging goal to accomplish if local residents and stakeholders are not *empowered* in matters of governance, to obtain, control and direct the use of goods and services towards broader sustainability oriented goals, such as long-term environmental conservation, community well-being, *etc.* A greater focus on "good governance" at the local level is therefore needed in the integrated SCBT approach.

Issues of equity, fairness and justice were found to be surprisingly under-represented in the literature review (one wonders why, as sustainability involves issues of fairness, historical injustices, trade-offs and distribution/use of often limited, non-renewable resources, *etc.*). We make a start to address this in Table 6 and Table 7. In order to emphasize the under-represented domain of governance and justice drawn from the literature, these are presented separately in Table 7below. They offer the potential to bridge the global discourse or ST and the local discourse of CBT, by bringing together universal and rights-based discourses (e.g., distributive justice, procedural justice) with local, particular issues (including historical injustices to specific groups), situated justice and an ethic of care and attention to diversity and "difference". In this sense, it can be argued that justice enables a crucial bridge between the global and the local, the universal and the particular. Issues of tourism development involve ethical issues of equity, fairness and justice in enabling global, planetary sustainability and local, community well-being.

5. WHERE DO WE GO FROM HERE? DIRECTIONS FORWARD

While preliminary, the notion of SCBT proposed in this article offers some important avenues to direct future research and practice in this area. The work summarized above reflects the history of ideas in the evolution of sustainable tourism and community-based tourism. It reveals a complex terrain of conflicting terms, concepts, principles and values and a vast number of criteria that have occupied researchers and practitioners. Yet, a close look at the information presented above offers strong justification for an integrated approach to ST and CBT, *i.e.*, sustainable community-based tourism. The greater focus on environmental sustainability that the concept of sustainable tourism inherited from its roots in sustainable development has been slowly offset with increasing attention to social and cultural impacts of tourism development and to the economic and overall well-being of local *communities*. The scale of sustainable tourism has evolved from global discourses (e.g., of UNWTO, UNEP [21]) to implementation by various organizations and destinations at the regional and local level worldwide, with policies emerging at the country-level in some cases, like Australia (e.g., [136,137]). Meanwhile, despite a strong focus on community development, growing awareness of the need to manage community commons, environmental health and conservation of natural goods is enabling a longer-term sustainability horizon to be incorporated into various CBT practices and forms (consider community-based ecotourism, for instance). Similar efforts have arisen with respect to the principle of *equity* in ST and CBT. Community-oriented tourism initiatives, like CBT, while presenting some similarities to sustainable tourism, also espouse some distinctly different, vital development goals (Table 5). As Saayman and Giampiccoli [138] put it, as an alternative to conventional mass tourism, CBT emphasizes control by community members, and the tourism-related benefits are intended to address economic disadvantage and social justice. Variants of such community-focused approaches, like pro-poor tourism (PPT), are specifically focused on redistributive aims and, as Butler, Curran and O'Gorman [56] show with their study of PPT in the regeneration project of Glasgow Govan (U.K.), are applicable to both the developed and developing world. Butler *et al.*[56] argue that government should be enforcing PPT principles rather than leaving this to businesses or other enterprises. The differences between ST and CBT thus begin to diminish in terms of geographic scope as examples arise of both forms around the world and with the inclusion of local resident communities as key stakeholders in ST [44,139]. This seems to be a good base from which to bridge the macro-micro, local-global scale and complexity of the tourism system, and Table 6 and Table 7 appear to be a reasonable starting point to commence the task of developing an SCBT framework to inform development, planning and management.

However, in going forward, there are some critical issues that our historical exploration above have just begun to reveal. Some repeated concerns and issues raised by previous researchers might be resolved through an integrated framework. For instance, Cardenas, Byrd and Duffy [140] summarize well about the importance of a good knowledge base among residents and other stakeholders who are invited to participate directly in tourism development, planning and management, facilitating important principles related to local control and the fair distribution of tourism resources and benefits. "Before informed, active, full, or meaningful participation can be achieved, tourism planners need to evaluate stakeholder level of awareness and perception of tourism, the tourism process, impacts, and principles of sustainability"(p. 254). They develop a tool called the Sustainable Tourism Development Index (SUSTDI) to assess the stakeholder (including community groups) awareness of tourism impacts and their perspectives on sustainable tourism development principles. Yet, despite all of the progress made over the last few decades in addressing crucial issues of environmental and social sustainability, individual, societal and industry well-being, concerns about the usefulness of the ST- and CBT-related concepts and failures of implementation abound across the globe remain. Three areas of pressing concerns are highlighted below, building on the history of ideas in sustainability-oriented and community-based approaches to tourism laid out above and which we believe are crucial to address if the proposed SCBT framework is to be successful. By itself, it too may result in inconsistent results and continued muddling through mazes of disparate views, values and approaches in tourism research and practice.

5.1. In the Grips of Neoliberal Globalization: The Primacy of the Local, the Failure of Academia

Bramwell and Lane [141] in their special issue on Governance in the *Journal of Sustainable Tourism* note that "tailored and effective governance" is crucial for implementing sustainable tourism, enhancing democratic processes, providing direction and facilitating practical progress. They argue that theoretical frameworks are crucial. One of the most strikingly obvious yet immensely puzzling observation that has arisen from our exploration thus far is not only the lack of theoretical guidance, but also the systematic failure to progress from critiques and calls for local control and community involvement in the literature, to a clear understanding and implementation of "good governance" in theory and practice. Ruhanen [142] undertakes a study of the transfer of academic knowledge on sustainability to tourism public sector practice in five destinations in Queensland, Australia. Her results indicate a knowledge-practice, gap, *i.e.*, that the knowledge based on this topic in academia has not been diffused effectively to planners and managers. Dredge and Jamal [143] adopt a poststructuralist approach to their study of planning and policy studies in tourism. Their findings lead them to conclude that "mainstream subjects related to destination development and management dominate while critical analysis of economic and political structures, interests and values is lagging" and tourism planning and policy studies must progress "towards greater visibility, legitimacy and importance in tourism studies through more critical engagement with tourism public policy and planning practice".

However, in almost three decades since the release of the Brundtland Commission's report on sustainable development, and more than that since the rise of RT and CBT, a "critical" engagement with the role of the public sector power, empowerment and control over the use and distribution of the public commons, long-term sustainability of environmental and social-cultural goods and planning and management practices have resulted in little shifting of power and "community-driven" tourism continues to occur in small pockets at best, with calls worldwide for greater integration and engagement of the public sector. Zou, Huang and Ding [144], for instance, forward a community-driven development (CDD) model in China for which they believe "theoretical implications could be derived from this paper to direct tourism planning exercises in other countries, especially those developing countries with similar economic development backgrounds to China" (p. 261). Their model is oriented toward three key factors: localization of the supply chain, community-external investor symbiosis and democratization of decision making. Zou *et al.*[144] observe that the "Manila Declaration on the Social Impact of Tourism"[139], "promises to maximize the positive effect of tourism through extensive community participation...Maintaining local control is also one of the important principles advocated by WTO in developing tourism" [144] (p. 263). Acknowledging that government reforms in China still had a long way to go, they believe the CDD model of rural tourism is an optimal way for sustainable rural tourism development in China and needs to be institutionalized and recognized in the government policy agenda. Their insights, like those of other academics addressing sustainability and tourism in and from post colonial democracies, as well as post-socialist, post-communist and post-apartheid contexts, speak to the importance of focusing greater attention on the role of the public sector, from the national to the local level (see, for example, [145,146]). In keeping with other proponents of poverty alleviation, Chili and Xulu [145] feel strongly that "government at all levels has the obligation to ensuring that the plight of the poor is addressed and turned around through sustainable tourism development" (p. 27). They, note, too that:

Overall, there limited literature that explores the role of local governments to facilitate and spearhead sustainable tourism development especially in developing countries (Yukdsel, Bramwell, and Yuksel [147]). In most cases governments tend to have numerous and promising policies and plans for sustainable tourism development which unfortunately do not yield good results because of deficiencies and shortcomings on execution and implementation.

[145] (p. 27)

The gap between academic research and practice and the continued lag in both planning and policy to instantiate the common accord in ST and CBT for local control, resident involvement and resident participation in tourism is troubling. However, even more worrisome is the lack of attention to the commonly agreed on principles of local control, oriented especially towards social justice and fair distribution of tourism goods and benefits to local residents, including diverse populations and disadvantaged groups. While a SCBT-based approach seems favorable to addressing the various spatial and temporal aspects noted above in the complex tourism system, how well can it address this continuing lag, which is especially worrisome in the present-day context of neoliberal globalization? Saayman and Giampiccoli [138] offer a succinct analysis of the current neoliberal milieu, concerning matters of poverty reduction and community development. As they note, within this context "specific issues of ideology, discourse, policies and related matters in general and their correlation to tourism need to be considered and sketched out" and, furthermore, "effort should be made to scale-up CBT concepts and practices to have greater impact. There is no conceptual or practical restriction to the scale-up of CBT development. Effective global restructuring of the tourism system cannot be on small scale" (p. 164). Their argument for scaling up from the micro-local to the macro-global level bodes well for an integrated approach to SCBT, but the continued failure of tourism studies to tackle tendentious political and ideological issues related to tourism policy and planning is highly problematic and deserving of much closer scrutiny and efforts to engage in academic praxis in the face of a global neoliberalization of local-global environmental and social systems. Dredge and Jamal [143] reiterate this call for greater academic attention, especially in light of neoliberalist shifts from public administration to *public management* over the last few decades:

...from the 1980s onwards, neoliberalism, globalisation and new public management have prompted a downsizing and outsourcing of government functions and a move away from direct government involvement in economic and social affairs. The role of government has been recast as a facilitator and enabler of economic activity...This shift is described as a move from public administration to public management, and has been characterised by the increasing uptake of public-private partnerships, collaborative planning and policy development and government- business power sharing (Bramwell, and Hall [98,148]). It has also meant that governments' relationship with public interests has become increasingly blurred... the notion of an overarching set of collective public interests for a broad and encompassing public good has been progressively abandoned by government policy makers in favour of a neoliberalist view....

[143] (p. 287)

5.2. What Are "Good" Principles for SCBT?

Table 6 shows an integrated set of criteria that are commonly cited under the well-established dimensions of environmental, social and economic impacts in the comprehensive literature review that we undertook. Table 7 shows some criteria, the additional dimensions of governance, justice and ethics, which are integral to development and sustainability goals. We use criteria here intentionally, as this was the most common item used in ST and CBT studies, from which we are able to compare and note similarities and differences between the two concepts. Closer attention is needed to clarify and reduce ambiguities in future research and to ground ST and CBT in a clear framework of ethical and management principles; both kinds are cited, but management principles in the context of environmental, societal and individual sustainability (in addition to economic and business sustainability) are grounded in ethical principles, though they may not be evident. The field of tourism studies in the 21st century can no longer ignore the global, but interrelated issues of environmental degradation, human disadvantage, poverty alleviation and climate change and clear ethical principles to guide research and practices are imperative. Developing such a framework of ethical principles for an integrated approach to the development and management of tourism at the local-global level is of vital importance. By bringing ST and CBT together as SCBT, it may be easier to undertake the task of grounding the complex tourism system in ethical principles that inform good governance, good practice, good management and good tourism (see Levy and Hawkins [149]).

It should be noted here that, as mentioned earlier, the array of concepts, terms, criteria, indicators and principles cited in the literature have tended to overlap or be confused in the meaning of each term. The Scopus literature review revealed a significant gap with respect to identifying key ST and CBT principles used by researchers. Other than commonly-known ones, such as noted earlier in the article (e.g., sustainable tourism development should be equitable in meeting the needs of current and future generations of tourists and residents; local communities should have control over tourism decision making in CBT; or a PPT principle that tourism benefits should be distributed to the most disadvantaged in the community), the principles are inconsistent , vague or hidden within general criteria, which are then used as "principles" (see, for example, Helmy [150], who aims to use ST principles in the research, but then switches to using ST "criteria" later).

What constitutes "good tourism" seems to be enmeshed in a swath of constructs with little agreement about these among academics or practitioners, and philosophical or theoretical guidance has been highly limited. Moscardo and Murphy [151] argue that "[w]hile the management of tourism impacts and the relationship between tourism and sustainability have been paid considerable attention by tourism academics, there is little evidence of any significant change in tourism practice" (p. 2538). To them, it reflects problems in the way tourism academics have conceptualized sustainable tourism, and they offer an alternative framework that addresses "quality-of-life, recognizes the complexity of tourism within local and global systems, adheres to the principles of responsible tourism, and explicitly assesses the value of tourism as one tool, amongst many, for sustainability" (p. 2538). Yet, what principles and capacities help to enable "quality of life" can vary significantly, and theory building is desperately needed to address moral questions about the good life and good tourism that could help ground sustainability principles, environmental and community well-being and individual quality of life in relation to tourism [104,152,153]. A limited number of studies have arisen to evaluate the success of sustainability in tourism-related endeavors using planning and management tools that draw upon key ethical theories (see, e.g., [92,154]).

5.3. SCBT Guided by Justice and an Ethic of Care?

Whitford and Ruhanen's [137] study of Australian State/Territory governments' policy for indigenous tourism examines how sustainable development principles are addressed and finds a "sustainability rhetoric" with little success in achieving sustainable tourism development for indigenous peoples. Based only on this study, the authors state that "one size fits all" framework for indigenous tourism development cannot work, and "policies need to draw upon indigenous diversity and, in a consistent, collaborative, coordinated and integrated manner, provide the mechanisms and capacity-building to facilitate long-term sustainable indigenous tourism" (p. 492). Little is mentioned of the historical context of indigenous oppression and possible structural conditions that continue to embed possible discrimination and racism, historically, in the Australian post-colony. Manyara and Jones [59] feel that the failure of CBEs in Kenya is related to foreign resource control and heavy reliance on donor funding that instantiate dependency and neocolonialism. However, how effective will their call be to integrate principles of sustainable development into community tourism development in Kenya, empower local leaders and communities and implement appropriate policy frameworks, if structural and institutional conditions that facilitate dependency and underdevelopment are not addressed? This calls not simply for radical inquiry into political issues and areas that, with some exceptions, scholars in tourism studies appear reluctant to tackle. Critical tourism studies have emerged as a young field of inquiry, and a small, but growing base of critique is emerging on topics such as neoliberalism, political economy, political ecology and "sustainability" in tourism (see, for example, [88,155,156]).

In an era of neoliberal globalization and severe challenges of climate change, tourism researchers and practitioners are being increasingly faced with issues of local and global justice, historically-driven inequities, structurally-embedded discrimination and institutional racism. Justice-related matters are just being taken up in tourism studies (see the studies mentioned in the previous section related to environmental and social justice), but much needs to be done in the context of sustainability and community-based tourism (SCBT in our case). Jamal and Camargo [109] forward a perspective on "destination justice" that calls to supplement Enlightenment- and modernity-driven notions of distributive justice used by Western liberal democracies (see, for instance, Rawls [117]) with an "ethic of care" drawing from Aristotelean and feminist perspectives for a more situated justice based on virtue ethics and care.

6. CONCLUSIONS

Tracing the progress of sustainable tourism and community-based tourism, two parallel domains of research and practice, is a challenging endeavor. This study demonstrates the confusing arrays of names and terminologies, goals and success factors, as well as conflating of principles, criteria, factors, *etc.* Not surprisingly, a number of criticisms and critics have arisen along the way: For example, Moscardo [157] argues that there is no such thing as "sustainable tourism." She concludes that it would be better to simply view tourism as one among many possible development options to aid in socioeconomic development in the destination region, which might help facilitate innovation, development of a different set of evaluation criteria and "the development, and more extensive use, of sustainability monitoring systems" [157] (p. 9). However, what ethical principles guide the development of the evaluation criteria and sustainability monitoring systems? What constitutes "good tourism" or "good tourism governance"? We argue that, rather than abandoning the notion of ST and CBT, the ethical principles that guide tourism development and governance need be made clear and applicable from an integrated, local to global perspective. Bridging the local (CBT) and the global (ST) are principles of good governance and justice that enable fairness and equity in the distribution and use of tourism-related resources from the local to the global level, as well as principles of community empowerment and capacity building, stewardship of natural, cultural and social goods.

While much remains to be done, this study, the very preliminary framework proposed and the discussion in the previous section are intended to offer some constructive hope and new directions for research and practice in ST and CBT. Our analysis of these knowledge domains calls for greater attention to developing an integrated, local-global approach to sustainable, community-based tourism development, where good governance aided by clearly-defined ethical and justice principles help to bridge the local-global "divide" to guide tourism development and management. In the era of neoliberal globalization today, many scholars worry that attention to social justice and related issues of gender inequity, class and power at the local level, as well as the global objectives of inter- and intra-generational equity and reducing inequalities between the north and the south through tourism will falter without active attention to the imperative of "good governance" [158,159]. Active intervention to facilitate knowledge transfer, collaborative engagement between academia, government and tourism practitioners worldwide to facilitate the priorities outlined above is a critical role that we, as students and researchers from diverse academic backgrounds and regions worldwide, can no longer shirk from and must speak to each other about.

ACKNOWLEDGMENTS

The open access publishing fees for this article have been covered by the Texas A&M University Online Access to Knowledge Fund (OAKFund), supported by the University Libraries and the Office of the Vice President for Research. We also wish to thank the organizers of the 23rd Nordic Symposium on Tourism and Hospitality (NORTHORS) conference, University of Copenhagen, Copenhagen, Denmark, 2–4 October 2014, for the opportunity to present an early draft of this manuscript.

AUTHOR CONTRIBUTIONS

Both of the authors contributed equally to the design, development and analysis of the proposed SCBT approach and the entire research manuscript.

CONFLICTS OF INTEREST

The authors declare no conflicts of interest.

REFERENCES

1. Jamal, T.; Camargo, B.A.; Wilson, E. Critical omissions and new directions for sustainable tourism: A situated macro-micro approach. *Sustainability* **2013**, *5*, 4594–4613. [Google Scholar] [CrossRef]
2. Wheeller, B. Tourism's troubled times: Responsible tourism is not the answer. *Tour. Manag.* **1991**, *12*, 91–96. [Google Scholar] [CrossRef]
3. Garrod, B.; Fyall, A. Beyond the rhetoric of sustainable tourism? *Tour. Manag.* **1998**, *19*, 199–212. [Google Scholar] [CrossRef]
4. Butler, R.W. Problems and issues of integrating tourism development. In *Contemporary Issues in Tourism Development*; Pearce, D.G., Butler, R.W., Eds.; Routledge: New York, NY, USA, 1999; pp. 65–80. [Google Scholar]
5. Graci, S.; Dodds, R. *Sustainable Tourism in Island Destinations*; Earthscan: Washington, DC, USA, 2010; pp. 1–225. [Google Scholar]
6. Goodwin, H. *Taking Responsibility for Tourism*; Goodfellow Publishers Limited: Woodeaton, UK, 2011; pp. 1–256. [Google Scholar]
7. Weaver, D.B. Organic, incremental and induced paths to sustainable mass tourism convergence. *Tour. Manag.* **2012**, *33*, 1030–1037. [Google Scholar] [CrossRef]

8. Springett, D.; Redcliff, M. Sustainable development: History and evolution of the concept. In*Rutledge International Handbook of Sustainable Development*; Redclift, M., Springett, D., Eds.; Routledge: New York, NY, USA, 2015; pp. 3–38. [Google Scholar]

9. Arksey, H.; O'Malley, L. Scoping studies: Towards a methodological framework. *Int. J. Soc. Res. Meth.* **2005**, *8*, 19–32. [Google Scholar] [CrossRef]

10. Grant, M.J.; Booth, A. A typology of reviews: An analysis of 14 review types and associated methodologies. *Health Inf. Libraries J.* **2009**, *26*, 91–108. [Google Scholar] [CrossRef] [PubMed]

11. Smith, V.L. Alternative/responsible tourism seminar. *Ann. Tour. Res.* **1990**, *17*, 479–480. [Google Scholar] [CrossRef]

12. United Nations Environment Program (UNEP). Declaration of the United Nations Conference on the Human Environment. Available online: http://www.unep.org/Documents.Multilingual/Default.asp?documentid=97&articleid=1503 (accessed on 28 January 2016).

13. United Nations Educational, Scientific and Cultural Organization (UNESCO). About World Heritage. Available online: http://whc.unesco.org/en/conventiontext/ (accessed on 28 January 2016).

14. Hall, C.M.; Gossling, S.; Scott, D. Tourism and sustainability: Towards a green(er) tourism economy? In *The Routledge Handbook of Tourism and Sustainability*; Hall, C.M., Gossling, S., Scott, D., Eds.; Routledge: New York, NY, USA, 2015; pp. 490–519. [Google Scholar]

15. United Nations Environment Program (UNEP). Global Environmental Outlook. Available online: http://www.unep.org/geo/GEO3/english/049.htm (accessed on 28 January 2016).

16. United Nations World Commission on Environment and Development (UNWECD). *Our Common Future*; Oxford University Press: New York, NY, USA, 1987. [Google Scholar]

17. De Kadt, E. *Tourism: Passport to Development? Perspectives on the Social and Cultural Effects of Tourism in Developing Countries*; Oxford University Press: New York, NY, USA, 1979. [Google Scholar]

18. Carruthers, D. From opposition to orthodoxy: The remaking of sustainable development. In*Debating the Earth: The Environmental Politics Reader*, 2nd ed.; Dryzek, J.S, Scholsberg, D., Eds.; Oxford University Press: New York, NY, USA, 2005; pp. 285–300. [Google Scholar]

19. World Travel and Tourism Council, United Nations World Tourism Organization and the Earth Council (WTTC, UNWTO, & EC). Agenda 21 for Travel and Tourism Industry: Towards Environmentally Sustainable Development, A Report, 1995. Available online: http://wwwv1.agora21.org/johannesburg/rapports/omt-a21.html (accessed on 28 January 2016).

20. United Nations World Tourism Organization (UNWTO). *Agenda 21 for Travel and Tourism: Towards Environmentally Sustainable Tourism*; WTO, WTTC and the Earth Council: London, UK, 1994. [Google Scholar]

21. United Nations Environment Program, (United Nations) World Tourism Organization (UNEP-UNWTO, 2005). *Making Tourism more Sustainable-A Guide for Policy Makers*; UNWTO: Madrid, Spain, 2005; pp. 9 and 20. Available online: http://www.unep.fr/shared/publications/pdf/DTIx0592xPA-TourismPolicyEN.pdf (accessed on 28 January 2016). [Google Scholar]

22. United Nations. Towards a Sustainable Development Agenda. Available online: http://www.un.org/sustainabledevelopment/development-agenda/ (accessed on 20 July 2015).

23. Hall, C.M.; Gossling, S.; Scott, D. *The Routledge Handbook of Tourism and Sustainability*; Routledge: New York, NY, USA, 2015. [Google Scholar]

24. Pisano, U.; Endl, A.; Berger, G. ESDN-2012 Report. Available online: http://www.sd-network.eu/quarterly%20reports/report%20files/pdf/2012-June-The_Rio+20_Conference_2012.pdf(accessed on 10 December 2015).

25. United Nations Development Program (UNDP), Sustainable Development Goals. Available online: http://www.undp.org/content/undp/en/home/mdgoverview/post-2015-development-agenda.html (accessed on 10 December 2015).

26. United Nations World Tourism Organization (UNWTO–GCET). Global Code of Ethics for Tourism. Available online: http://ethics.unwto.org/en/content/global-code-ethics-tourism(accessed on 10 December 2015).

27. Global Sustainable Tourism Council (GSTC) Criteria for destinations. Available online: http://www.gstcouncil.org/ (accessed on 7 May 2016).

28. Bricker, K.S.; Schultz, J. Sustainable tourism in the USA: A comparative look at the Global Sustainable Tourism Criteria. *Tour. Recreat. Res.* **2011**, *36*, 215–229. [Google Scholar] [CrossRef]

29. Mowforth, M.; Munt, I. *Tourism and Sustainability: New Tourism in the Third World*; Routledge: New York, NY, USA, 1998. [Google Scholar]

30. Wall, G. Sustainable tourism—Unsustainable development. In *Tourism Development and Growth*; Wahab, S., Pigram, J.J., Eds.; Routledge: New York, NY, USA, 1997; pp. 33–49. [Google Scholar]

31. Ecumenical Coalition on Tourism (ECOT). Available online: http://www.tourism-watch.de/en/content/25-years-ecumenical-coalition-tourism (accessed on 10 February 2016).

32. Blamey, R.K. Principles of ecotourism. In *The Encyclopedia of Ecotourism*; Weaver, D.B., Ed.; CABI Publishing: Cambridge, MA, USA, 2001; pp. 5–22. [Google Scholar]

33. Fennell, D.A. A content analysis of ecotourism definitions. *Curr. Issues Tour.* **2001**, *4*, 403–421. [Google Scholar] [CrossRef]

34. Honey, M. *Ecotourism and Sustainable Development: Who Owns Paradise?* Island Press: Washington, DC, USA, 2008. [Google Scholar]

35. The International Center for Responsible Tourism (ICRT). Available online: http://www.icrtourism.org/ (accessed on 10 March 2015).

36. Miller, G.; Twining-Ward, L. *Monitoring for a Sustainable Tourism Transition: The Challenge of Developing and Using Indicators*; CABI Publishing: Cambridge, MA, USA, 2005. [Google Scholar]

37. Goodwin, H. Indigenous tourism and poverty reduction. In *Tourism and Indigenous Peoples: Issues and Implication*; Butler, R., Hinch, T., Eds.; Elsevier: Philadelphia, PA, USA, 2007. [Google Scholar]

38. Goodwin, H. Pro-poor tourism: A response. *Third World Q.* **2008**, *29*, 869–871. [Google Scholar] [CrossRef]

39. Goodwin, H. *Taking Responsibility for Tourism*; Goodfellow Publishers Ltd.: Oxford, UK, 2008. [Google Scholar]

40. Goodwin, H. Tourism, local economic development, and poverty reduction. *Appl. Res. Econ. Dev.* **2008**, *5*, 55–64. [Google Scholar]

41. Scheyvens, R. Ecotourism and the empowerment of local communities. *Tour. Manag.* **1999**,*20*, 245–249. [Google Scholar] [CrossRef]

42. Scheyvens, R. *Tourism for Development: Empowering Communities*; Prentice-Hall: Harlow, UK, 2002. [Google Scholar]

43. Lucchetti, V.G.; Font, X. Community based tourism: Critical success factors. Available online:www.icrtourism.org/wp-content/uploads/2012/03/OP27.pdf (accessed on 10 February 2016).

44. Bramwell, B.; Lane, B. Sustainable tourism: An evolving global approach. *J. Sustain. Tour.* **1993**, *1*, 1–5. [Google Scholar] [CrossRef]

45. Hunter, C.J. On the need to re-conceptualise sustainable tourism development. *J. Sustain. Tour.* **1995**, *3*, 155–165. [Google Scholar] [CrossRef]

46. Hardy, A.; Beeton, R.J.; Pearson, L. Sustainable tourism: An overview of the concept and its position in relation to conceptualizations of tourism. *J. Sustain. Tour.* **2002**, *10*, 475–496. [Google Scholar] [CrossRef]

47. Sharpley, R. Tourism and sustainable development: Exploring the theoretical divide. *J. Sustain. Tour.* **2000**, *8*, 1–19. [Google Scholar] [CrossRef]

48. Farrell, B.; Twining-Ward, L. Seven steps towards sustainability: Tourism in the context of new knowledge. *J. Sustain. Tour.* **2005**, *13*, 109–122. [Google Scholar] [CrossRef]

49. Dredge, D., Jenkins, J., Eds.; *Tourism Planning and Policy*; John Wiley: Milton, ON, Canada, 2007.

50. Jamal, T.B.; Getz, D. Collaboration theory and community tourism planning. *Ann. Tour. Res.* **1995**, *22*, 186–204. [Google Scholar] [CrossRef]

51. Redcliff, M., Springett, D., Eds.; *Routledge International Handbook of Sustainable Development*; Earthscan: New York, NY, USA, 2015.

52. Weaver, D. Community-based tourism as strategic dead-end. *Tour. Recreat. Res.* **2010**, *35*, 206–208. [Google Scholar] [CrossRef]

53. Murphy, P.E. *Tourism: A Community Approach*; Methuen: New York, NY, USA, 1985. [Google Scholar]

54. Ellis, S.; Sheridan, L.M. The legacy of war for community-based tourism development: Learnings from Cambodia. *Commun. Dev. J.* **2014**, *49*, 129–142. [Google Scholar] [CrossRef]

55. Hall, C.M. *Introduction to Tourism in Australia: Impacts, Planning and Development*; Addison Wesley Longman: Melbourne, Australia, 1996. [Google Scholar]

56. Butler, R.; Curran, R.; O'Gorman, K.D. Pro-poor tourism in a first world urban setting: Case study of Glasgow Govan. *Int. J. Tour. Res.* **2013**, *15*, 443–457. [Google Scholar] [CrossRef]

57. APEC Tourism Working Group. *Effective Community Based Tourism: A Best Practice Manual*; Asia-Pacific Economic Cooperation: Sydney, Australia, 2010. [Google Scholar]

58. Netherlands Development Organization (SNV) Asia Pro-Poor Sustainable Tourism Network.*A Toolkit for Monitoring and Managing Community-Based Tourism*; SNV Asia Pro-Poor Sustainable Tourism Network and Griffith University: Queensland, Australia, 2007. [Google Scholar]

59. Manyara, G.; Jones, E. Community-based tourism enterprises development in Kenya: An exploration of their potential as avenues of poverty reduction. *J. Sustain. Tour.* **2007**, *15*, 628–644. [Google Scholar] [CrossRef]

60. Cole, S. Information and empowerment: The keys to achieving sustainable tourism. *J. Sustain. Tour.* **2006**, *14*, 629–644. [Google Scholar] [CrossRef]

61. Vajirakachorn, T. Determinants of Success for Community-Based Tourism: The Case of Floating Markets in Thailand. Ph.D. Thesis, Texas A&M University, College Station, TX, USA, 2011. [Google Scholar]

62. Hatton, M.J. *Community-Based Tourism in the Asia-Pacific for Canadian Tourism Commission, APEC, and Canadian International Development Agency*; School of Media Studies at Humber College: Toronto, ON, Canada, 1999. [Google Scholar]

63. Billington, R.D.; Carter, N.; Kayamba, L. The practical application of sustainable tourism development principles: A case study of

creating innovative place-making tourism strategies. *Tour. Hosp. Res.* **2008**, *8*, 37–43. [Google Scholar] [CrossRef]

64. Matarrita-Cascante, D. Beyond growth, reaching tourism-led development. *Ann. Tour. Res.* **2010**, *37*, 1141–1163. [Google Scholar] [CrossRef]

65. The International Ecotourism Society (TIES). About–What Is Ecotourism? Available online: http://www.ecotourism.org/what-is-ecotourism (accessed on 15 March 2015).

66. Mbaiwa, J.E.; Stronza, A.L. The effects of tourism development on rural livelihoods in the Okavango Delta, Botswana. *J. Sustain. Tour.* **2010**, *18*, 635–656. [Google Scholar] [CrossRef]

67. Strambach, S.; Surmeier, A. Knowledge dynamics in setting sustainable standards in tourism–the case of 'Fair Trade in Tourism South Africa'. *Curr. Issues Tour.* **2013**, *16*, 736–752. [Google Scholar] [CrossRef]

68. Fair Trade Tourism Annual Report 2012/2013. Available online: http://www.fairtrade.travel/uploads/files/2012–2013_Annual_Report.pdf (accessed on 18 February 2016).

69. Simmons, D.G. Community participation in tourism planning. *Tour. Manag.* **1994**, *15*, 98–108. [Google Scholar] [CrossRef]

70. Sharpley, R. Flagship attractions and sustainable rural tourism development: The case of the Alnwick Garden, UK. *J. Sustain. Tour.* **2007**, *15*, 125–143. [Google Scholar] [CrossRef]

71. Poitras, L.; Donald, G. Sustainable wine tourism: The host community perspective. *J. Sustain. Tour.* **2006**, *14*, 425–448. [Google Scholar] [CrossRef]

72. Li, Y. Exploring community tourism in China: The case of Nanshan cultural tourism zone. *J. Sustain. Tour.* **2004**, *12*, 175–193. [Google Scholar] [CrossRef]

73. Tosun, C. Limits to community participation in the tourism development process in developing countries. *Tour. Manag.* **2000**, *21*, 613–633. [Google Scholar] [CrossRef]

74. Mansuri, G.; Rao, V. Community-based and-driven development: A critical review. *World Bank Res. Obs.* **2004**, *19*, 1–39. [Google Scholar] [CrossRef]

75. Boley, B.B.; McGehee, N.G.; Perdue, R.R.; Long, P. Empowerment and resident attitudes toward tourism: Strengthening the theoretical foundation through a Weberian lens. *Ann. Tour. Res.* **2014**, *49*, 33–50. [Google Scholar] [CrossRef]

76. United Nations Environment Program (UNEP). A Three Year Journey for Sustainable Tourism. Available online: http://www.unep.org/resourceefficiency/Portals/24147/scp/tourism/activities/taskforce/pdf/TF%20REPORT_FINAL.pdf (accessed on 9 February 2016).

77. Ashley, C.; Roe, D.; Goodwin, H. Pro-Poor Tourism Strategies: Making Tourism Work for the Poor, A review of experience, Pro-poor Tourism Report 1, ICRT, IIED and ODI. Available online: http://www.odi.org/sites/odi.org.uk/files/odi-assets/publications-opinion-files/3246.pdf (accessed on 26 January 2015).

78. Russell, P. Community-based tourism. *Travel Tour. Anal.* **2000**, *5*, 89–116. [Google Scholar]

79. Jamal, T.; Stronza, A. Collaboration theory and tourism practice in protected areas: Stakeholders, structuring and sustainability. *J. Sustainable Tour.* **2009**, *17*, 169–189. [Google Scholar] [CrossRef]

80. Arnstein, S.R. A ladder of citizen participation. *J. Am. Inst. Plan.* **1969**, *35*, 216–224. [Google Scholar] [CrossRef]

81. Peeters, P. A clear path towards sustainable mass tourism? Rejoinder to the paper 'Organic, incremental and induced paths to sustainable mass tourism convergence' by David, B. Weaver. *Tour. Manag.* **2012**, *33*, 1038–1041. [Google Scholar] [CrossRef]

82. Dredge, D.; Jamal, T. Mobilities on the Gold Coast, Australia: Implications for destination governance and sustainable tourism. *J. Sustain. Tour.* **2013**, *21*, 557–579. [Google Scholar] [CrossRef]

83. Moscardo, G. Exploring social representations of tourism planning: Issues for governance. *J. Sustain. Tour.* **2011**, *19*, 423–436. [Google Scholar] [CrossRef]

84. Mahanti, A.; Manuel-Navarrete, D. From sustainable development to governance for sustainability. In *Routledge International Handbook of Sustainable Development*; Redclift, M., Springett, D., Eds.; Routledge: New York, NY, USA, 2015; pp. 416–428. [Google Scholar]

85. Gössling, S., Hall, C.M., Weaver, D., Eds.; *Sustainable Tourism Futures: Perspectives on Systems, Restructuring and Innovations*; Routledge: New York, NY, USA, 2009.

86. Dodds, R.; Ali, A.; Galaski, K. Mobilizing knowledge: Determining key elements for success and pitfalls in developing community-based tourism. *Curr. Issues Tour.* **2016**, 1–22. [Google Scholar] [CrossRef]

87. Chettiparamb, A.; Kokkranikal, J. Responsible tourism and sustainability: The case of Kumarakom in Kerala, India. *J. Polic Res. Tour. Leis. Events* **2012**, *4*, 302–326. [Google Scholar] [CrossRef]

88. Britton, S.G. The political economy of tourism in the Third World. *Ann. Tour. Res.* **1982**, *9*, 331–358. [Google Scholar] [CrossRef]

89. Wood, R.E. Tourism and international policy: Neoliberalism and beyond. In *The SAGE Handbook of Tourism Studies*; Jamal, T., Robinson, M., Eds.; Sage: Newcastle upon Tyne, UK, 2009. [Google Scholar]

90. Johnston, T. Sustainable tourism development. In *Routledge International Handbook of Sustainable Development*; Redclift, M., Springett, D., Eds.; Routledge: New York, NY, USA, 2015; pp. 250–263. [Google Scholar]

91. Jamal, T.; Borges, M.; Stronza, A. The institutionalisation of ecotourism: Certification, cultural equity and praxis. *J. Ecotour.* **2006**, *5*, 145–175. [Google Scholar] [CrossRef]

92. García-Rosell, J.C.; Mäkinen, J. An integrative framework for sustainability evaluation in tourism: Applying the framework to tourism product development in Finnish Lapland. *J. Sustain. Tour.* **2013**, *21*, 396–416. [Google Scholar] [CrossRef]

93. Farsari, I. The development of a conceptual model to support sustainable tourism policy in north Mediterranean destinations. *J. Hosp. Mark. Manag.* **2012**, *21*, 710–738. [Google Scholar] [CrossRef]

94. Stoddard, J.E.; Pollard, C.E.; Evans, M.R. The triple bottom line: A framework for sustainable tourism development. *Int. J. Hosp. Tour. Adm.* **2012**, *13*, 233–258. [Google Scholar] [CrossRef]

95. Siow May, L.; Abidin, Z.Z.; Nair, V.; Ramachandran, S.; Shuib, A. Developing criteria and indicators for responsible rural tourism

in Taman Negara National Park (TNNP), Malaysia. *Malays. For.* **2011**, *74*, 143–156. [Google Scholar]

96. Roberts, S.; Tribe, J. Sustainability indicators for small tourism enterprises–An exploratory perspective. *J. Sustain. Tour.* **2008**, *16*, 575–594. [Google Scholar] [CrossRef]

97. Puhakka, R.; Sarkki, S.; Cottrell, S.P.; Siikamäki, P. Local discourses and international initiatives: Sociocultural sustainability of tourism in Oulanka National Park, Finland. *J. Sustain. Tour.* **2009**, *17*, 529–549. [Google Scholar] [CrossRef]

98. Bramwell, B. Governance, the state and sustainable tourism: A political economy approach. *J. Sustain. Tour.* **2011**, *19*, 459–477. [Google Scholar] [CrossRef]

99. Barke, M.; Towner, J. Learning from experience? Progress towards a sustainable future for tourism in the Central and Eastern Andalucian Littoral. *J. Sustain. Tour.* **2003**, *11*, 162–180. [Google Scholar] [CrossRef]

100. Pomering, A.; Noble, G.; Johnson, L.W. Conceptualising a contemporary marketing mix for sustainable tourism. *J. Sustain. Tour.* **2011**, *19*, 953–969. [Google Scholar] [CrossRef]

101. García-Melón, M.; Gómez-Navarro, T.; Acuña-Dutra, S. A combined ANP-delphi approach to evaluate sustainable tourism. *Environ. Impact Assess. Rev.* **2012**, *34*, 41–50. [Google Scholar] [CrossRef]

102. Hultsman, J. Just tourism: An ethical framework. *Ann. Tour. Res.* **1995**, *22*, 553–567. [Google Scholar] [CrossRef]

103. Tribe, J. Education for ethical tourism action. *J. Sustain. Tour.* **2002**, *10*, 309–324. [Google Scholar] [CrossRef]

104. Smith, M.; Duffy, R. *The Ethics of Tourism Development*; Routledge: New York, NY, USA, 2003. [Google Scholar]

105. Macbeth, J. Towards an ethics platform for tourism. *Ann. Tour. Res.* **2005**, *32*, 962–984. [Google Scholar] [CrossRef]

106. Higgins-Desbiolles, F. Justice tourism and alternative globalisation. *J. Sustain. Tour.* **2008**, *16*, 345–364. [Google Scholar] [CrossRef]

107. Higgins-Desbiolles, F. Justifying tourism: Justice through tourism. In *Tourism and Inequality: Problems and Prospects*; Cole, S., Morgan, N., Eds.; University of Wales Institute: Cardiff, Wales, 2010; pp. 194–240. [Google Scholar]

108. Lee, S.; Jamal, T. Environmental justice and environmental equity in tourism: Missing links to sustainability. *J. Ecotour.* **2008**, *7*, 44–67. [Google Scholar] [CrossRef]

109. Jamal, T.; Camargo, B.A. Sustainable tourism, justice and an ethic of care: Toward the just destination. *J. Sustain. Tour.* **2014**, *22*, 11–30. [Google Scholar] [CrossRef]

110. Richins, H. Environmental, cultural, economic and socio-community sustainability: A framework for sustainable tourism in resort destinations. *Environ. Dev. Sustain.* **2009**, *11*, 785–800. [Google Scholar] [CrossRef]

111. Catibog-Sinha, C.; Wen, J. Sustainable tourism planning and management model for protected natural areas: Xishuangbanna biosphere reserve, South China. *Asia Pac. J. Tour. Res.* **2008**, *13*, 145–162. [Google Scholar] [CrossRef]

112. Ellis, S.; Sheridan, L.A. Critical reflection on the role of stakeholders in sustainable tourism development in least-developed countries. *Tour. Plan. Dev.* **2014**, *11*, 467–471. [Google Scholar] [CrossRef]

113. Wight, P.A. Supporting the principles of sustainable development in tourism and ecotourism: Government's potential role. *Curr. Issues Tour.* **2002**, *5*, 222–244. [Google Scholar] [CrossRef]

114. Ottenbacher, M.C.; Schwebler, S.; Metzler, D.; Harrington, R.J. Sustainability Criteria for Tourism Attractions: A Case Study of Germany. *Int. J. Soc. Ecol. Sustain. Dev. (IJSESD)* **2015**,*6*, 20–39. [Google Scholar] [CrossRef]

115. United Nations World Tourism Organizations (UNWTO). *Guide for Local Authorities on Developing Sustainable Tourism*; UNWTO: Madrid, Spain, 1998. [Google Scholar]

116. Mohonk Agreement. Proposal for an International Certification Program for Sustainable Tourism and Ecotourism. Available online: http://www.rainforest-alliance.org/tourism/documents/mohonk.pdf (accessed on 15 December 2015).

117. Rawls, J. *A Theory of Justice*; The Belknap Press of Harvard University Press: Cambridge, MA, USA, 1999. [Google Scholar]

118. Fennell, D. Ethics and tourism. In *Philosophical Issues in Tourism*; Tribe, J., Ed.; Channel View Publications: Bristol, UK, 2009; pp. 211–242. [Google Scholar]

119. Holden, A. *Environment and Tourism*; Routledge: New York, NY, USA, 2000. [Google Scholar]

120. Sharpley, R., Telfer, D.J., Eds.; *Tourism and Development: Concepts and Issues*, 2nd ed.; Channel View Publications: Bristol, UK, 2015.

121. Butcher, J. Against ethical tourism. In *Philosophical Issues in Tourism*; Tribe, J., Ed.; Channel View Publications: Bristol, UK, 2009; pp. 244–277. [Google Scholar]

122. United Nations World Tourism Organization (UNWTO). *Indicators of Sustainable Development for Tourism Destinations: A Guidebook*; UNWTO: Madrid, Spain, 2004. [Google Scholar]

123. Tanguay, G.A.; Rajaonson, J.; Therrien, M.C. Sustainable tourism indicators: Selection criteria for policy implementation and scientific recognition. *J. Sustain. Tour.* **2013**, *21*, 862–879. [Google Scholar] [CrossRef]

124. Choi, H.C.; Sirakaya, E. Sustainability indicators for managing community tourism. *Tour. Manag.* **2006**, *27*, 1274–1289. [Google Scholar] [CrossRef]

125. Spenceley, A. Nature-based tourism and environmental sustainability in South Africa. *J. Sustain. Tour.* **2005**, *13*, 136–170. [Google Scholar] [CrossRef]

126. Schianetz, K.; Kavanagh, L. Sustainability indicators for tourism destinations: A complex adaptive systems approach using systemic indicator systems. *J. Sustain. Tour.* **2008**, *16*, 601–628.cultural he. [Google Scholar] [CrossRef]

127. Choo, H.; Jamal, T. Tourism on organic farms in South Korea: A new form of ecotourism. *J. Sustain. Tour.* **2009**, *17*, 431–454. [Google Scholar] [CrossRef]

128. Park, D.B.; Yoon, Y.S. Developing sustainable rural tourism evaluation indicators. *Int. J. Tour. Res.* **2011**, *13*, 401–415. [Google Scholar] [CrossRef]

129. Reddy, M.V. Sustainable tourism rapid indicators for less-developed islands: an economic perspective. *Int. J. Tour. Res.* **2008**, *10*, 557–576. [Google Scholar] [CrossRef]

130. Hall, C.M. Policy learning and policy failure in sustainable tourism governance: from first-and second-order to third-order change. *J. Sustain. Tour.* **2011**, *19*, 649–671. [Google Scholar] [CrossRef]

131. Wray, M. Adopting and implementing a transactive approach to sustainable tourism planning: Translating theory into practice. *J. Sustain. Tour.* **2011**, *19*, 605–627. [Google Scholar] [CrossRef]

132. Eagles, P.F.; Romagosa, F.; Buteau-Duitschaever, W.C.; Havitz, M.; Glover, T.D.; McCutcheon, B. Good governance in protected areas: An evaluation of stakeholders' perceptions in British Columbia and Ontario Provincial Parks. *J. Sustain. Tour.* **2013**, *21*, 60–79. [Google Scholar] [CrossRef]

133. Jamal, T.; Watt, E.M. Climate change pedagogy and performative action: Toward community-based destination governance. *J. Sustain. Tour.* **2011**, *19*, 571–588. [Google Scholar] [CrossRef]

134. Beaumont, N.; Dredge, D. Local tourism governance: A comparison of three network approaches. *J. Sustain. Tour.* **2010**, *18*, 7–28. [Google Scholar] [CrossRef]

135. Peterson, T.R. *Sharing the Earth: The Rhetoric of Sustainable Development*; University of South Carolina Press: Columbia, SC, USA, 1997. [Google Scholar]

136. Muangasame, K.; McKercher, B. The challenge of implementing sustainable tourism policy: A 360-degree assessment of Thailand's "7 Greens sustainable tourism policy". *J. Sustain. Tour.* **2015**, *23*, 497–516. [Google Scholar] [CrossRef]

137. Whitford, M.M.; Ruhanen, L.M. Australian indigenous tourism policy: Practical and sustainable policies? *J. Sustain. Tour.* **2010**, *18*, 475–496. [Google Scholar] [CrossRef]

138. Saayman, M.; Giampiccoli, A. Community-based and pro-poor tourism: Initial assessment of their relation to community development. *Eur. J. Tour. Res.* **2016**, *12*, 145. [Google Scholar]

139. World Tourism Organization (WTO). *Manila Declaration on the Social Impact of Tourism*; WTO: Manila, Philippines, 1997. [Google Scholar]

140. Cárdenas, D.A.; Byrd, E.T.; Duffy, L.N. An exploratory study of community awareness of impacts and agreement to sustainable tourism development principles. *Tour. Hosp. Res.***2015**, *15*, 254–266. [Google Scholar] [CrossRef]

141. Bramwell, B.; Lane, B. Critical research on the governance of tourism and sustainability. *J. Sustain. Tour.* **2011**, *19*, 411–421. [Google Scholar] [CrossRef]

142. Ruhanen, L. Progressing the sustainability debate: A knowledge management approach to sustainable tourism planning. *Curr. Issues Tour.* **2008**, *11*, 429–455. [Google Scholar] [CrossRef]

143. Dredge, D.; Jamal, T. Progress in tourism planning and policy: A post-structural perspective on knowledge production. *Tour. Manag.* **2015**, *51*, 285–297. [Google Scholar] [CrossRef]

144. Zou, T.; Huang, S.S.; Ding, P. Toward a community-driven development model of rural tourism: The Chinese experience. *Int. J. Tour. Res.* **2014**, *16*, 261–271. [Google Scholar] [CrossRef]

145. Chili, N.S.; Xulu, N. The role of local government to facilitate and spearhead sustainable tourism development. *Probl. Perspect. Manag.* **2015**, *13*, 27–31. [Google Scholar]

146. Grundey, D. Managing sustainable tourism in Lithuania: Dream or reality? *Technol. Econm. Dev. Econ.* **2008**, *14*, 118–129. [Google Scholar] [CrossRef]

147. Yukdsel, F.; Bramwell, B.; Yuksel, A. Centralized and decentralized tourism governance in Turkey. *Ann. Tour. Res.* **2005**, *32*, 859–878. [Google Scholar] [CrossRef]

148. Hall, C.M. A typology of governance and its implications for tourism policy analysis. *J. Sustain. Tour.* **2011**, *19*, 437–457. [Google Scholar] [CrossRef]

149. Levy, S.E.; Hawkins, D.E. Peace through tourism: Commerce based principles and practices.*J. Bus. Ethics* **2009**, *89*, 569–585. [Google Scholar] [CrossRef]

150. Helmy, E. Towards integration of sustainability into tourism planning in developing countries: Egypt as a case study. *Curr. Issues Tour.* **2004**, *7*, 478–501. [Google Scholar] [CrossRef]

151. Moscardo, G.; Murphy, L. There is no such thing as sustainable tourism: Re-conceptualizing tourism as a tool for sustainability. *Sustainability* **2014**, *6*, 2538–2561. [Google Scholar] [CrossRef]

152. Jamal, T.; Menzel, C. Good actions in tourism. In *Philosophical Issues in Tourism*; Tribe, J., Ed.; Channel View Publications: Bristol, UK, 2009; pp. 227–243. [Google Scholar]

153. Fennell, D.A. *Tourism Ethics*; Channel View Publications: Bristol, UK, 2006. [Google Scholar]

154. Briedenhann, J. Socio-cultural criteria for the evaluation of rural tourism projects–a Delphi consultation. *Curr. Issues Tour.* **2009**, *12*, 379–396. [Google Scholar] [CrossRef]

155. Telfer, D.J. Development studies and tourism. In *The SAGE Handbook of Tourism Studies*; Jamal, T., Robinson, M., Eds.; Sage: Newcastle upon Tyne, UK, 2009. [Google Scholar]

156. Mowforth, M.; Munt, I. *Tourism and Sustainability: Development, Globalisation and New Tourism in the Third World*, 3rd ed.; Routledge: New York, NY, USA, 2009. [Google Scholar]

157. Moscardo, G. Sustainable tourism innovation: Challenging basic assumptions. *Tour. Hosp. Res.* **2008**, *8*, 4–13. [Google Scholar] [CrossRef]

158. Camus, S.; Hikkerova, L.; Sahut, J.M. Systemic analysis and model of sustainable tourism.*Int. J. Bus.* **2012**, *17*, 365. [Google Scholar]

159. Duffy, L.N.; Mowatt, R.A.; Cnacellor, H.C.; Cárdenas, D.A. Machismo–marianismo and the involvement of women in a community-based tourism project in Ecuador, South America.*Tour. Anal.* **2012**, *17*, 791–803.

3

CHAPTER

Local Residents' Attitude toward Sustainable Rural Tourism Development

Iulia C. Muresan [1,*], Camelia F. Oroian [1], Rezhen Harun [2,*], Felix H. Arion [1,*], Andra Porutiu [1], Gabriela O. Chiciudean [1], Alexandru Todea [1] and Ramona Lile [3]

[1]Department of Economic Sciences, University of Agricultural Sciences and Veterinary Medicine Cluj-Napoca, 3-5 Manastur Street, Cluj-Napoca 400372, Romania

[2]Department of Agribusiness and Rural Development, Faculty of Agricultural Sciences, University of Sulaimani, Kurdistan Regional Government-Iraq, Sulaimani-Bakrajo 5100, Iraq

[3]Department of Economics, Faculty of Economic Sciences, "Aurel Vlaicu" University of Arad, 77 Bdul. Revolutiei, Arad 310130, Romania

ABSTRACT

: Tourism is a multi-faced activity that links the economic, social and environmental components of sustainability. This research analyzes rural residents' perceptions of the impact of tourism development and examines the factors that influence the support for sustainable tourism development in the region of Nord-Vest in Romania. Residents' perceptions towards tourism development were measured using 22 items, while their support for tourism development was determined using 8 items. Descriptive and inferential statistics were used to analyze the data. Principal component analysis grouped the first 22 variables into 4 factors, and the following 8 variables into 2 factors (sustainable development, destination development). Findings indicate that residents see tourism as a development factor. The natural, economic, and social-cultural environment as well as infrastructure, age, gender and education are factors that influence the sustainable development of tourism.

Keywords: rural tourism; tourism planning; residents' attitudes; tourism impact

1. INTRODUCTION

The importance of tourism nowadays is determined by the multiple roles that it plays within any country (economic, social and cultural) and its ability to create a positive impact (employment, wealth, dynamism, income enhancement, infrastructure, international friendship and moving people and assets) [1,2]. The tourism sector has become, during the last several decades, a major factor of importance in world GDP because of its rapid growth and development [3,4], often associated with an export industry [5]. Eshliki and Kaboudi [6] describe the tourism sector as a "powerful force of change in the economy," mainly due to its positive economic impact on communities. Stetic [7] underlines the key role that rural areas play in the tourism sector, not only because of their special position as areas of excellence, but also as ecological oases combined with the ability to preserve traditional culture and ethno-cultural heritage. Therefore, Vazques *et al.* [8] appreciate the importance of rural tourism as a key factor for proper socio-economic development, while Giannakis [9] highlights the importance of rural areas within the European Union from the point of view of their vast territory held (91%) and the population involved (59%). These rural areas can be strengthened and revitalized only through proper and viable tourism based on sustainable development [10]. Because of the high importance of the tourism sector, the newest Tourism Action Framework at the European level mentions among its main priorities the stimulation of long-term competitiveness and the promotion of sustainable development [11]. The main advantages of the sustainable approach in tourism are related to good practices and improvements, especially regarding the environment; however, at the same time, the concepts' incapacity for a proper implementation in practice has been observed [12]. Either way, sustainable tourism development is strongly related to local communities and their attitudes towards tourism. Within the process of developing a sustainable tourism sector, the local community is the key element, as it is directly affected by its evolution [6].

Development of tourism influences the development of other sectors such as agriculture, food processing and handcrafting, which can contribute to the well-being of the entire community. Previous studies about Romanian rural tourism analyzed the factors that influence rural tourism development [13,14], the particularities of rural tourism potential [15,16]], supply [17,18] and demand [17,19]. Local residents represent an important component of the amalgam that constitutes the destination [20] and influences the future development of any tourism destination.

The aim of this paper is to identify the support of the local community towards sustainable tourism development in the region of Nord-Vest in Romania. To this end, residents' perceptions towards the impact of tourism development are identified, and, secondly, the residents' support for future tourism development is determined. The analysis reveals that the support for future tourism development is based on two components: sustainable development and destination development. In this context, the paper presents an original approach towards the subject, the results of which can be useful in creating future development strategies.

2. LITERATURE REVIEW

The idea of sustainable tourism development emerged in the last several decades as a necessity to ensure an efficient tourism sector based on three main components: environmental interests, socio-cultural and economic needs of the communities involved [21]. To maintain a balance between the positive and negative impacts that the three factors could generate, one must take into consideration the local community—the core element within the tourism development process and the most important stakeholder [22,23,24]. Analyzing the local community's perceptions regarding tourism impact becomes a major concern, because it is strongly connected to the will to support tourism development [25,26]. Generally, tourism is perceived by residents as having strong economic benefits, which outweigh any other possible negative impacts, encouraging residents to perceive tourist activity in a positive way and resulting in strong involvement and support [26,27]. Still, interesting findings were noted in Vietnam, where touristic activity is supported by the community not for its economic benefits, but rather for its socio-cultural and environmental benefits [28].

Studies focus both on the negative and positive impacts of tourism on local communities. For example, communities in Egypt have been shown to support tourist activity, even if both negative and positive impacts are perceived, because the positive are stronger than the negative [3]. In terms of economic development, positive impacts are reflected in job creation, investments and increasing the national income, while negative impacts are related to an increased level of inflation. Socio-cultural factors can generate two kinds of attitudes: positive ones related to cultural exchange, and negative ones linked to the way of life and overcrowded places [3]. In research conducted in Arizona, Mcgehee and Andereck [27] concluded that support for tourism development is highly related to personal benefits obtained by the local community members, where overall the community perceives tourism in a positive way due to its effects on the local economy (such as job creation and an improvement in quality of life). A study conducted in Mauritius observed a similar positive economic impact due to job creation and improvements in quality of life. A positive relationship was observed between the support for tourism and economic and socio-cultural factors, while a negative relationship was established between community support and the environment [5].

It has been observed that the main components of sustainable tourism development influence the local community's goodwill and support for tourism; therefore, much research focuses on studying its economic, environmental and socio-cultural impacts [3,5,22,27,29,30]. Even if tourism offers important positive benefits, its strong development could cause major dissatisfaction among communities due to intensive traffic, inflation, and crime [6,29].

The relationship between community satisfaction and tourism development analyzed by Min *et al.* (2012) [22] shows that, even if the level of satisfaction related to tourist activity in the area is not high, the community still supports it when it is considered promising for future city development. The residents' support for tourist activity is connected also to their perceptions of economic, environmental and socio-cultural factors. Positive impact on aforementioned factors indicate high support from the local community [23,30,31].

An important finding was obtained by Koa and Stewart [29] after conducting a study in the Keju Islands of Korea, a relatively undeveloped area. Results indicated that residents' attitudes toward tourism are directly related to the stage of development of the host community. In Uganda it was observed that the local community has a positive attitude towards tourism because it is considered as a factor of development generating incomes, increasing agricultural production and "good fortune" [32]. The results of research conducted in North Carolina indicated two main reasons for which the community is willing to support tourist activities, which are strongly related to the personal benefits obtained: The female population positively perceives the cultural dimension represented by the development of arts, craft and household items, while the youngsters perceive improvements in social life and recreation facilities [33]. The community attachment and involvement, as well as the support for sustainable tourism development, are influenced by their perceived benefits to local residents [34].

Understanding the residents' perceptions towards tourism is important in order to shape future policies that minimize the potential negative impact of tourism and maximize its benefits [35].

3. MATERIALS AND METHODS

The study was conducted in a rural area in the region of Nord-Vest in Romania from November 2014 to April 2015. Nord-Vest comprises 6 counties: Bihor, Bistrita-Nasaud, Cluj, Maramures, Satu-Mare and Salaj. It has a surface area of 34,159 km^2, representing 14.32% of the total country surface. The region borders both Hungary and Ukraine [36].

This region has a high potential for tourism due to its natural environment and landscape (mountains, natural reservations, thermal waters, salt mines) and to a variety of cultural tourist attractions (churches, wooden churches, traditions, *etc.*). The region includes 170 protected areas of national importance, of which two are national parks (Rodna Mountains National Park and Calimani National Park) and two natural parks (Apuseni Natural Park and Maramures Mountains Natural Park) [37]. The parks, both national and natural, attract tourists due to the possibilities of hiking, bird watching, rural tourism and agritourism [38,39].

Table 1 presents a timeline of tourism's supply and demand from 2005 to 2014. In Nord-Vest, at the end of 2014, there were 676 (11% of the total number at national level) lodging facilities, of which 33% (226) were agritourism guesthouses, which represent the main type of accommodation facility in the rural area. During the last 10 years in Nord-Vest, the number of lodgings rose by 40% and the number of agritourism guesthouses 61%, while the occupancy rate and the average length of stay remained quite similar. This can be explained by the increase in the number of tourist arrivals and overnight stays. The arrivals in agritourism guesthouses in the region of Nord-Vest increased by almost 200%, and the overnight stays by 188% from 2005 to 2014. The number of guesthouses increased from 2005 to 2010 by 90% mainly due to European funds that supported rural development, and by the end of 2014 they decreased by 15%. Factors which could have led to this decreasing trend may include managers' lack of specific entrepreneurial education and skills [40], and owners' lack of capacity to adapt and diversify their services according to the needs of the tourists [41]. The majority of the guesthouses assure basic facilities (bed & breakfast); in this case, it can be stated that Romanian rural tourism lacks additional touristic services for entertainment, outdoor and indoor recreation, handicrafts and other souvenirs [42].

Table 1. Evolution of tourist activity indicators in Romania and North-West Region.

Region	Indicator	Year									
		2005	2006	2007	2008	2009	2010	2011	2012	2013	2014
Romania	Number of lodgings	4226	4710	4694	4840	5095	5222	5003	5821	6009	6130
	Number guest houses	956	1259	1292	1348	1412	1354	1210	1569	1598	1665
	Total average length of stay (nights) *	3.16	3.06	2.95	2.91	2.82	2.64	2.56	2.49	2.44	2.40
	Guesthouses average length of stay (nights) *	2.15	2.12	2.05	2.08	2.07	2.09	2.06	2.03	1.99	1.97
	Total occupancy rate (%) *	33.42	33.61	36.04	35.02	28.35	25.16	26.28	25.85	25.14	26.11
	Guest houses occupancy rate (%) *	14.47	14.41	16.34	18.41	14.22	12.36	13.78	13.20	12.56	13.16
Northwest	Number of lodgings	480	543	554	585	645	658	650	730	709	676
	Number guest houses	140	193	200	225	259	266	206	252	231	226
	Total average length of stay (nights) *	3.12	3.03	2.87	2.79	2.87	2.68	2.61	2.48	2.35	2.35
	Guesthouses average length of stay (nights) *	1.99	1.98	1.92	1.95	1.93	1.98	2.04	2.03	1.92	1.92
	Total occupancy rate (%)*	32.24	32.06	34.05	32.69	27.70	23.25	23.61	22.33	22.58	25.38
	Guest houses occupancy rate (%) *	13.16	15.17	19.74	24.22	14.52	9.35	9.77	10.76	10.95	13.15

Source: National Institute of Statistics, Tempo-online time-series [43]; * compute based on data from National Institute of Statistics.

To examine residents' perceptions towards the impact of tourism development in Nord-Vest, the authors used a quantitative survey. The collected data can be divided into three main sections: perception of tourism impact, support of tourism development and socio-demographic characteristics of the respondents. The 22 variables used to determine the rural residents' perceptions of tourism development were adopted from previous studies [3,5,6,27,29,33,44,45,46,47]. The support for tourism development was measured with 8 variables developed by [47,48]. A 5-point Likert-type scale was used based on the following scale: 1 = strongly disagree, 2 = disagree, 3 = neutral, 4 = agree, 5 = strongly agree to evaluate each variable.

Data were collected from 433 residents from the rural area of Nord-Vest. The sample size was determined based on the following formula [30].

$$n = \frac{(1.96)^2 \times 1075725 \times 0.5 \times (1 - 0.5)}{(0.05)^2 \times (1075725 - 1) + (1.96)^2 \times 0.5 \times (1 - 0.5)} = 400\text{residents} \quad (1)$$

550 questionnaires were distributed to a proportional stratified random sample of adults. The response rate was 91% (502 questionnaires), and in the end 433 questionnaires were validated.

The selection of the respondents was based on their age respective of the distribution of the original population.

Descriptive statistics were used to determine the socio-demographic profile of the rural residents and to describe the rural residents' perceptions of tourism development as well as their support for it. Exploratory factor analysis was used to assess the factor structure of the variables. Principal component analysis (PCA) is a data reduction method in which the components are calculated using all of the variance of the manifest variables, with all of that variance appearing in the solution [49]. This is achieved by transforming a new set of variables, which are uncorrelated, and which are ordered so that the first few explain most of the variance [50]. Mathematically, this is equivalent to finding the best low rank approximation of the data via the singular value decomposition [51]. Two principal component analyses were conducted separately, the first to group the perception variables about tourism development, and the second to group the variables regarding future tourism development support. The varimax rotation was used to maximize the differences among the components extracted and to maintain correlation among the components. A simple correlation analysis (Spearman rank-order correlation coefficient) was utilized to calculate the correlation between the support for sustainable tourism development and the impact of tourism, and between the support for tourism destination development and the impact of tourism development.

4. RESULTS AND DISCUSSION

4.1. Rural Residents' Characteristics

The socio-demographic characteristics of the respondents are presented in Table 2. According to the National Institute of Statistics, at the end of the year 2014, 57% of people from rural areas were more than 40 years old. This data is reaffirmed by the results of the current study. The results can mainly be attributed to the lack of attractive employment opportunities in the rural area for the younger generation. Another problem in the rural area is the low level of education. Most of the respondents had graduated from high school (45.4%), while 28% had less than high school education. The monthly household income levels reported are less than 225 Euro (36.4%), 225–445 Euro (36%), more than 445 Euro (27%), while the average family number is 3.7 members. It can be concluded that the rural population was aging and less educated, with a low monthly income.

4.2. Factors of Rural Residents' Perception towards Tourism Development

Principal factor analysis was conducted to assess the dimensionality of the 22 items. The Barlett test of sphericity is significant (Chi-square = 3915.62, $p < 0.000$). The Kaiser-Meyer-Olkin (KMO) overall measure of sampling is 0.88, indicating that data are suitable for the principal component analysis [52]. Values of 0.6 or above from the KMO measures indicated that data are adequate for PCA [53]. The PCA with varimax rotation of the 22 variables resulted in a four-component solution that explains 57.66% of the total variance. Only factors with eigenvalues greater than one were selected [54]. Cronbach's alpha reliability coefficient was computed to evaluate the internal consistency of each component. An acceptable reliability coefficient is higher than 0.6 [54,55,56]. The overall reliability of the 22 variables was 0.87.

Table 2. Characteristics of the respondents.

Variables	%
Gender	
Female	41.57
Male	58.43
Age	
15–19	1.20
20–29	18.2
30–39	24.1
40–49	28.8
50–59	19.6
>60	8.00
Education	
Illiterate	0.20
Less than high school	28.1
High school	45.4
University	26.2
House hold income	
<225 euro	36.4
225–445 euro	35.9
>445 euro	27.7

The components recorded after the first principal component analysis are listed in Table 3. Component 1 comprises 8 variables (0.87 alpha), component 2 comprises 6 variables (0.82 alpha), component 3 comprises 5 variables (0.73 alpha), and component 4 comprises 3 variables (0.67 alpha).

Table 3. Principal component analysis on tourism impact variables.

Eigenvalue	Variance %	Component	Item	Factor Loading	Communalities
6.10	27.74	Environmental α = 0.87	Development of tourism damage natural environment and landscape	0.793	0.632
			Tourism cause overcrowding problems for residents	0.777	0.661
			Tourism increase the air pollution	0.775	0.641
			Tourists use too much water	0.770	0.598
			Tourism results in more litter in an area	0.739	0.571
			Tourism development negatively affects the recreational facilities and entertainment	0.690	0.537
			The construction of tourist facilities destroy the environment	0.674	0.461
			Increase traffic problems	0.561	0.490
3.67	16.68	Economical α = 0.82	Tourism plays an important role in the economic development of the area	0.824	0.696
			Tourism improves locals standard of living	0.802	0.735
			Tourism increases a community's tax revenue	0.699	0.557
			Tourism create new jobs for locals	0.649	0.562
			Tourism diversifies the rural economy	0.546	0.502
			Tourism results in an increase in the cost of living	0.540	0.624
			Revenue from tourism taxes activity should be invested in future development of tourism	0.523	0.386
1.88	8.56	Social and Cultural α = 0.73	Tourism provide incentives for restoration of traditional houses	0.781	0.639
			Interaction with tourists is a positive experience	0.776	0.617
			Shopping and restaurants option is better as a result of tourism	0.676	0.552
			Tourism development enhance more recreational opportunities for locals	0.431	0.468
1.03	4.67	Infrastructure α = 0.67	Improves traffic network	0.798	0.693
			Improves living utilities infrastructure (supply of water. sewage. electric etc)	0.711	0.583
			Quality of public services in better	0.508	0.480
Total variance %	57.66				

The first component labeled "environmental effects" explains 27.74% of the variance and has a mean of 2.58 (SD = 1.049). This component involves attributes that focus on conservation of natural resources and negative impacts of tourism on the environment. The environmental effects are seen as being the most negative for the rural residents in terms of tourism development, such as destroying the natural environment (factor loading 0.793), overcrowding problems (factor loading 0.777), air pollution (factor loading 0.775), and water scarcity

(0.770). The second component labeled "economic benefits" explains 16.68% of the variance and has a mean of 3.40 (SD = 0.913). This component involves attributes related to the overall economic development (factor loading 0.824), tax revenue (factor loading 0.699), employment (factor loading 0.649), and living cost (factor loading 0.540). The third component labeled "social and cultural impacts" involves attributes related to quality of life and cultural activities. It explains 8.56% of the variance and has a mean of 3.50 (SD = 1.008). This component groups items related to restoration of traditional houses (factor loading 0.781), diversification of recreational facilities (factor loading 0.431), alternative possibilities for shopping and dining (factor loading 0.676), and the perception of residents' interaction with tourists (factor loading 0.776). The fourth component explains 4.67% of the variance and involves attributes related to infrastructure development, being labeled "infrastructure benefits"; the mean is 3.19 (SD = 1.100) (Table 3 and Table 4). These results diverge from previous research, which has shown either that economic impacts may be more important [46,47], or that the social and cultural impacts were more significant [5] than the environmental component.

Table 4. Perception on tourism development impact.

Item	Mean	SD
Environment	2.58	1.049
Damage natural environment and landscape	2.45	1.393
Tourism cause overcrowding problems for residents	2.61	1.487
Tourism increase the air pollution	2.52	1.446
Tourists use too much water	2.33	1.398
Tourism results in more litter in an area	3.02	1.490
Tourism development negatively affects the recreational facilities and entertainment	2.60	1.463
The construction of tourist facilities destroy the environment	2.47	1.360
Increase traffic problems	2.69	1.412
Economic	3.40	0.913
Tourism plays an important role in the economic development of the area	3.78	1.235
Tourism improves locals standard of living	3.31	1.290
Tourism increases a community's tax revenue	3.73	1.294
Tourism create new jobs for locals	3.07	1.435
Tourism results in an increase in the cost of living	2.92	1.391
Tourism diversifies the rural economy	2.97	1.323
Revenue from tourism taxes activity should be invested in future development of tourism	4.09	1.158
Social and Cultural	3.5	1.008
Tourism provide incentives for restoration of traditional houses	3.35	1.488
Interaction with tourists is a positive experience	3.58	1.321
Shopping and restaurants option is better as a result of tourism	3.44	1.391
Tourism development enhance more recreational opportunities for locals	3.62	1.197
Physical	3.19	1.100
Improves traffic network	3.06	1.508
Improves living utilities infrastructure (supply of water, sewage, electric *etc.*)	3.19	1.459
Quality of public services in better	3.33	1.261

4.3. Local Residents' Perception towards Tourism Development Impacts

The host community tends to agree that development of tourism has a positive impact on the development of the region. The effects of tourism development on the natural environment and landscape are not perceived as being negative. The mean value of component 1 is 2.58, lower than the one found in a similar study [5]. The only variable with a higher score than 3 is the one related to the quantity of litter in the area.

The rural residents believe that the revenue from the taxation of touristic activity should be used for the future development of this sector (mean = 4.09), since an increase in the community's tax revenue is perceived (mean = 3.73). Tourism is seen as an important factor for the economic development of the area (mean = 3.78), which improves local standard of living (mean = 3.31) without affecting the cost of living (mean = 2.92). The findings support past studies regarding the people's perception towards the economic impact of tourism [27,46,47]. At the same time, it should be underlined that, for the rural residents of Nord-Vest, tourism is not perceived as an alternative to agricultural activities, nor as a factor of economic activities' diversification (mean = 2.97). One of the causes may be the lack of knowledge regarding the founding sources and the fear of business failure [40].

The rural residents agree that the development of tourism in their region provides more recreational opportunities (mean = 3.62), and interaction with the tourists is perceived as a positive experience (mean = 3.58). The rural residents perceive the development of tourism as a factor that provides cultural identity and improves the quality of services, and, indirectly, the standard of living. These results confirm the findings of previous studies [46,47]. Tourism leads to the restoration and preservation of the cultural values of the rural community (mean = 3.35). This is an important aspect for assuring sustainable development of the rural area.

Tourism is perceived as a factor that influences the development of the traffic network (mean = 3.06) and of the living utilities (mean = 3.19) and improves the quality of public services (mean = 3.33). Improvement in the quality of public services is a result of higher standards of the tourists who visit the area. Therefore, in order to satisfy the tourists' expectations, the services are adapted to their needs and quality standards are imposed.

4.4. Local Residents' Support for Future Tourism Development

Principal component analysis was used to assess the reliability of the 8 variables related to future sustainable tourism development. The Barlett test of sphericity is significant (Chi-square = 1026.348, $p < 0.000$). The Kaiser-Meyer-Olkin (KMO) overall measure of sampling is 0.84, indicating that data are suitable for the principal component analysis [52]. The PCA with varimax rotation of the 8 variables resulted in a two-component solution that explains 56.67% of the total variance. The overall reliability of the 8 variables is 0.83.

Four attitude variables ("plans are important to manage the growth of tourism," "long-term planning reduces the negative environmental impact," "authorities support tourism development," and "new environment protection measures") concerning the sustainable development of tourism were loaded in the first component with the cross-correlation coefficients of 0.820, 0.751, 0.747 and 0.703. This factor accounts for 42.89% of the total variance and was named support for tourism sustainable development (Table 5). The higher scores indicate that rural residents are focused on the long-term impact of tourism development.

The second component consists of 4 variables ("tourism represents a sustainable activity," "support for new facilities," "my community should become a tourist destination" and "tourism should become an important part of the community") with the cross-correlation coefficients of 0.756, 0.740, 0.665 and 0.514 (Table 5). This factor accounts 13.78% of the total variance and was termed support for tourism destination development.

The development of tourism should be supported by the local authorities (mean = 4.35) by setting strategic plans, with clear actions for tourism's growth management (mean = 4.11). The long-term planning with reduction of negative environmental impacts (mean = 3.76) and new environmental protection measures (mean = 3.95) are less supported than the involvement of local authorities and strategic planning of tourism development because of the reduced negative impact of tourism on the environment, as was noticed in Section 4.3.

The rural residents believe that tourism should be encouraged and become an important part of the community (mean = 4.12). This can be achieved by developing new tourism facilities (mean = 4.32) that will assure job alternatives for the local community. Furthermore, development plans should take into consideration the desire of the local residents for the sustainable development of their region as a tourism destination (mean = 4.12) (Table 6).

Table 5. Support for future tourism development.

Eigenvalue	Variance %	Component	Items	Factor Loading	Communalities
3.43	42.89	Sustainable development α = 0.77	It is important to develop plans to manage the growth of tourism	0.820	0.717
			Long-term planning will reduce the negative environmental impacts	0.751	0.565
			I agree that local authorities support tourism development	0.747	0.625
			New environment protection measures should be developed	0.703	0.544
1.10	13.78	Tourism destination α = 0.61	Tourism is a sustainable activity in my community	0.756	0.580
			I support new tourism facilities	0.740	0.617
			My community should become more of a tourist destination	0.665	0.617
			I support tourism and I would like to see it become an important part of my community	0.514	0.270
Total variance %	56.67				

Table 6. Degree of agreement for future tourism development.

Statements	Mean	SD
Sustainable development	4.03	0.961
It is important to develop plans to manage the growth of tourism	4.11	1.196
Long-term planning will reduce the negative environmental impacts	3.76	1.273
I agree that local authorities support tourism development	4.35	1.111
New environment protection measures should be developed	3.95	1.331
Destination development	4.24	0.820
I support new tourism facilities	4.32	0.994
Tourism is a sustainable activity in my community	4.12	1.134
My community should become more of a tourist destination	4.12	1.118
I support tourism and I would like to see it become an important part of my community	4.49	1.678

The results of simple correlation analysis on the support for sustainable tourism development and support for destination development for environmental impacts, economic benefits, socio-cultural impacts, infrastructure benefits, and age are listed in Table 7. The environmental impact ($r = -0.214, p < 0.01$), economic benefits ($r = 0.230, p < 0.01$), socio-cultural impacts ($r = 0.498, p < 0.01$) and infrastructure benefits ($r = 0.328, p < 0.01$) were significantly correlated with support for sustainable development. The environmental impact ($r = -0.252, p < 0.01$), economic benefits ($r = 0.241, p < 0.01$), socio-cultural impacts ($r = 0.418, p < 0.01$), and infrastructure benefits ($r = 0.292, p < 0.01$) were significantly correlated with support for future tourism destination development. Age does not have any influence on the residents' support for tourism development (Table 7). A *t*-Test was carried out to examine the influence of gender and education level on the support for future tourism development (Table 8 and Table 9).

Table 7. Correlation of each variable with the support for sustainable development and destination development.

Dependent Variable	Sustainable Development		Destination Development	
Independent Variable	Correlation Coefficient (r)	p-Value	Correlation Coefficient (r)	p-Value
Environment impact	−0.214 **	0.000	−0.252 **	0.000
Economic benefits	0.230 **	0.000	0.241 **	0.000
Social and cultural impacts	0.498 **	0.000	0.418 **	0.000
Infrastructure	0.328 **	0.000	0.292 **	0.000
Age	−0.036 *	0.457	−0.052	0.282

* significant at 0.05; ** significant at 0.01.

Table 8. Results of t-test analysis of gender and support for future tourism development.

Dependent Variable	Means		t Value	Sig.
	Female	Male		
Sustainable development	4.17	3.94	2.545	0.013 *
Destination development	4.34	4.18	2.017	0.044 *

* significant at 0.05; ** significant at 0.01.

Table 9. Results of t-test analysis of education level and support for future tourism development.

Dependent Variable	Means		t Value	Sig.
	Less than High School	More than High School		
Sustainable development	3.91	4.08	−1.696	0.091
Destination development	4.14	4.29	−1.715	0.087

* significant at 0.05; ** significant at 0.01.

There is an indirect link between perceived environmental impact and the support for sustainable development, and between environmental impact and support for tourism destination development. A direct link can be observed between economic impact, social and cultural impact, infrastructure benefits and the support of local community for future tourism development (Table 7).

Residents who perceive the impact of tourism less negatively tend to more highly support sustainable tourism development in the rural area. Referring to Table 8, the results show that females differ significantly in their support for tourism development (mean = 4.17 for sustainable development and mean = 4.34 for destination development). This can be explained by the fact that tourism represents an alternative activity for agriculture in the rural space, which can attract young people to establish themselves and work in the rural area. Furthermore, tourism represents a source of employment and entrepreneurial opportunities for women [57]. Results showed that residents who perceive the environmental impact of tourism more negatively tend to display diminished support for the future development of tourism, as was also suggested in previous studies [5,44].

Economic and socio-cultural benefits can be considered personal benefits of tourism development. The greater the benefits the rural residents perceive from tourism, the more likely they are to support sustainable tourism development, and the more likely they are to transform their community into a tourism destination. The findings support past studies regarding residents' support of tourism development, which found that the higher the personal benefits from tourism are, the more willing the local residents are to develop tourism in their community [28,33].

5. CONCLUSIONS

The research objectives were to investigate rural residents' attitudes towards various tourism impact variables and to explore their influence on the support of future sustainable tourism development. The results of this research indicate that rural residents perceive tourism development positively. Tourism is a sustainable development activity in the rural community. The environmental component of sustainable development is the most important one, a fact that can be explained by the high natural tourist potential and the awareness of the local residents of the importance of natural conservation for sustainable development, on the one hand, and the desire to reduce the negative effects of tourism development on the area on the other.

Tourism development improves the quality of life of local residents due to its effect on economic development of the area, which in turn leads to new employment opportunities. Furthermore, tourist activity in the rural area is perceived as being beneficial to the diversification of recreational alternatives and the improvement of the general infrastructure. Similar findings in examination of residents' attitudes toward tourism development were observed by Abdollahzadeh and Sharifzadeh [46].

Results indicate that tourism impacts are perceived positively as employment opportunity and well-being increases. An important segment of the rural population is willing to support the development of sustainable tourism because of the personal benefits obtained in terms of socio-cultural aspects (arts and crafts development, improvement of social life and facilities).

The results show that rural residents see tourism as an income generator, but at the same time they understand the importance of planning and managing tourism destinations sustainably. The local community, as an important stakeholder within the tourism sector, becomes a key element in developing future tourism strategies. Hanafiah [47] underlined that the community should get actively involved in the process of tourism development.

For rural residents, tourism is seen as an opportunity to enhance the wellbeing of the community in general, and particularly their own. The local community is willing to support sustainable tourism development if the personal benefits perceived are important. The greater the perceived economical, socio-cultural and infrastructure benefits, the higher the support is from the local community in building future tourism strategies.

This study revealed a lack of knowledge regarding residents' perceptions towards tourism impact and attitudes toward sustainable tourism development in a rural area of Romania. The results provide tourism planners and policymakers with viable information regarding rural residents' attitudes toward the development of tourism, which can serve as a useful tool in future development plans, enhancing sustainable tourism development and reducing tourism's negative impacts. Even if at the moment the residents tend to support sustainable tourism development, it is recommended that long-term changes in residents' attitudes and perceptions be taken into account. Furthermore, it was shown that the support for sustainable tourism development is greater if policymakers and tourism planners are attentive to residents' concerns and beliefs [28].

Over time, a lack of consistent and reliable information regarding local residents' attitudes towards sustainable rural development has negatively influenced the decision-making process regarding the sector's funding allocation. In fact, The National Rural Development Plan 2014–2020 of Romania includes two different specific measures indirectly related to rural tourism: Measure 06—"Development of exploitations and companies" (Under-measure 6.4 "Support for investments in creating and developing non-agricultural activities"), and Measure 07—"Basic services and village renewal in rural areas" (Under-measure 7.6 "Investments related to cultural heritage protection") [58]. No specific measure was dedicated to rural tourism during this period. In the first two years of implementation (2014 and 2015), 52 proposals were submitted (75,987,178 Euro) under Under-measure 6.4, while only 2 were accepted and financed (350,858 Euro), and none were financed in the first two years. Concerning Under-measure 7.6, 224 proposals were submitted (42,565,798 Euro), and not one was accepted and financed until the end of 2015 [59]. The results of this research can contribute to an improved strategy regarding rural tourism development, such as considering residents' requirements and expectations when creating the specifications of the under-measures, a process that would result in more sustainable rural tourism development.

Finally, this study has several limitations: primarily, limited time and low budget. Like its predecessors [27], this study did not clarify how the residents perceive themselves as benefiting from sustainable tourism development. Due to the variety of natural and anthropic tourism potential at this moment, based on the results of the research, it is impossible to formulate specific recommendations for different types of tourism products. Furthermore, the studies that follow should focus more on the particularities of the natural characteristics of the research area (thermal water, natural parks proximity, *etc.*), while more factors influencing the support for sustainable tourism development should be added (field of activity, land ownership, dependence on tourism, preoccupation for nature protection and responsible behavior) to determine the support for sustainable tourism development. In this way, important information can be gained and geared towards specific actions for future tourism development.

ACKNOWLEDGMENTS

Assistance provided by Molly McDonough, graduate of Master Food Identity, during the writing and English-proofing of the manuscript is greatly appreciated.

AUTHOR CONTRIBUTIONS

Iulia C. Muresan, Rezhen Harun and Felix H. Arion designed the research; Camelia F. Oroian, Alexandru Todea and Ramona Lile applied the survey; Camelia F. Oroian and Andra Porutiu organized the data based; Iulia C. Muresan and Gabriela O. Chiciudean analyzed the data, Iulia C. Muresan, Gabriela O. Chiciudean, Felix H. Arion and Rezhen Harun wrote the manuscript.

CONFLICTS OF INTEREST

The authors declare no conflict of interest.

REFERENCES

1. Egbali, N.; Nosrat, A.B.; Alipour, S.K.S. Effects of positive and negative rural tourism (Case study: Rural Semnan Province). *J. Geogr. Reg. Plan.* **2011**, *4*, 63–76. [Google Scholar]

2. Shariff, N.M.; Abidin, A.Z. Community attitude towards tourism impacts: Developing a standard instrument in the Malaysian context. *E-J. Soc. Sci. Res.* **2013**, *1*, 386–396. [Google Scholar]

3. Eraqi, M.I. Local communities' attitudes towards impacts of tourism development in Egypt.*Tour. Anal.* **2007**, *12*, 191–200. [Google Scholar] [CrossRef]

4. Budeanu, A. Impacts and responsibilities for sustainable tourism: A tour operator's perspective. *J. Clean. Product.* **2005**, *13*, 89–97. [Google Scholar] [CrossRef]

5. Ramseook-Munhurrun, P.; Naidoo, P. Residents' attitudes toward perceived tourism benefits. *Int. J. Manag. Mark. Res.* **2011**, *4*, 45–56. [Google Scholar]

6. Eshliki, S.A.; Kaboudi, M. Perception of community in tourism impacts and their participation in tourism planning: Ramsar, Iran. *J. Asian Behav. Stud.* **2012**, *2*, 51–64. [Google Scholar] [CrossRef]

7. Stetic, S. Specific features of rural tourism destinations management. *J. Settl. Spat. Plan.* **2012**,*1*, 131–137. [Google Scholar]

8. De la Torre, G.M.V.; Gutiérrez, E.M.A.; Guzman, T.J.L.G. Tourism as generator of wealth in rural areas. In Proceedings of the 18th European Advanced Studies Institute in Regional Science, Lodz-Cracow, Poland, 1–10 July 2005.

9. Giannakis, E. The role of rural tourism on the development of rural areas: The case of Cyprus Elias. *Romanian J. Reg. Sci.* **2014**, *8*, 38–53. [Google Scholar]

10. Garau, C. Perspectives on cultural and sustainable rural tourism in a smart region: The case study of Marmilla in Sardinia

(Italy). *Sustainability* **2015**, *7*, 6412–6434. [Google Scholar] [CrossRef]

11. Iunius, R.F.; Cismaru, L.D. Raising competitiveness for tourist destinations through information technologies within the newest tourism action framework proposed by the European commission. *Sustainability* **2015**, *7*, 12891–12909. [Google Scholar] [CrossRef]

12. Mihalic, T. Sustainable-responsible tourism discourse e towards 'responsustable' tourism sustainability. *J. Clean. Prod.* **2014**. [Google Scholar] [CrossRef]

13. Turnock, D. Sustainable rural tourism in the Romanian Charpatians. *Geogr. J.* **1999**, *165*, 192–199. [Google Scholar] [CrossRef]

14. Dragulanescu, I.V.; Drutu (Ivan), M. Rural tourism for local economic development. *Int. J. Acad. Res. Account. Financ. Manag. Sci.* **2012**, *2*, 196–203. [Google Scholar]

15. Iorio, M.; Corsale, A. Rural tourism and livelihood strategies in Romania. *J. Rural Stud.* **2010**,*26*, 152–162. [Google Scholar] [CrossRef]

16. Negrusa, A.L.; Cosma, S.A.; Bota, M. Romanian rural tourism development: A study case in Maramures. *Int. J. Bus. Res.* **2007**, *7*, 129–135. [Google Scholar]

17. Marin, D.; Petroman, C.; Petroman, I.; Ciolac, R.; Balan, I. Study regarding rural guest-house and agritourist household's number and percent in the total number of tourists establishments in Romania. *Lucr. Stiintifice Ser. Agron.* **2009**, *52*, 469–470. [Google Scholar]

18. Arion, F.; Muresan, I. An overview of rural tourism in the development regions of Romania.*Bull. UASVM Hortic.* **2008**, *62*, 35–40. [Google Scholar]

19. Naghiu, A.; Vasquez, J.L.; Georgiev, I. Rural development strategies through rural tourism activities in Romania: Chance for an international demand? *Int. Rev. Public Nonprofit Mark.***2005**, *2*, 85–95. [Google Scholar] [CrossRef]

20. Vanhove, N. *The Economics of Tourism Destination*; Elsevier Butterworth-Heinemann: Oxford, UK, 2005. [Google Scholar]

21. Mansfeld, Y.; Jonas, A. Evaluating the socio-cultural carrying capacity of rural tourism communities: A 'value stretch'

approach. *Tijdschr. Voor Econ. Soc. Geogr.* **2006**, *97*, 583–601. [Google Scholar] [CrossRef]

22. Min, Z.; Xiaoli, P.; Bihu, W. Research on residents' perceptions on tourism impacts and attitudes: A case study of Pingyao ancient city. In Proceedings of the 6th Conference of the International Forum on Urbanism (IFoU), Tourbanism, Barcelona, 25–27 January 2012; pp. 1–10.

23. Mohammadi, M.; Khalifah, Z.; Hosseini, H. Local people perceptions toward social, economic and environmental impacts of tourism in Kermanshah (Iran). *Asian Soc. Sci.* **2010**,*6*, 220–225. [Google Scholar] [CrossRef]

24. Banki, M.B.; Ismail, H.N. Multi-stakeholder perception of tourism impacts and ways tourism should be sustainably developed in obudu mountain resort. *Dev. Ctry. Stud.* **2014**, *4*, 37–48. [Google Scholar]

25. Huh, C.; Vogt, C.A. Changes in residents' attitudes toward tourism over time: A cohort analytical approach. *J. Travel Res.* **2008**, *46*, 446–455. [Google Scholar] [CrossRef]

26. Bestard, A.B.; Nadal, R.J. Attitudes toward tourism and tourism congestion. *Reg. Dev.* **2007**,*25*, 193–207. [Google Scholar]

27. McGehee, N.G.; Andereck, K.L. Factors predicting rural residents' support of tourism. *J. Travel Res.* **2004**, *43*, 131–140. [Google Scholar] [CrossRef]

28. Long, P.H. Perceptions of tourism impact and tourism development among residents of Cuc Phuong National Park, Ninh Binh, Victnam. *J. Ritsumeikan Soc. Sci. Hum.* **2011**, *3*, 75–92. [Google Scholar]

29. Ko, D.W.; Stewart, W.P. A structural equation model of residents' attitudes for tourism development. *Tour. Manag.* **2002**, *23*, 521–530. [Google Scholar] [CrossRef]

30. Brida, J.G.; Disegna, M.; Osti, L. Residents' perceptions of tourism impacts and attitudes towards tourism policies in a small mountain community. In Proceedings of the Ninth Canadian Congress on Leisure Research, Wolfville, NS, Canada, 12–15 May 1999.

31. Choi, H.C.; Murray, I. Resident attitudes toward sustainable community tourism. *J. Sustain. Tour.* **2010**, *18*, 575–594. [Google Scholar] [CrossRef]

32. Lepp, A. Residents' attitudes towards tourism in Bigodi village, Uganda Case study. *Tour. Manag.* **2007**, *28*, 876–885. [Google Scholar] [CrossRef]

33. Wang, Y.; Pfister, R.E. Residents attitudes toward tourism and perceived personal benefits in a rural community. *J. Travel Res.* **2008**, *47*, 84–93. [Google Scholar] [CrossRef]

34. Lee, T.H. Influence analysis of community resident support for sustainable tourism development. *Tour. Manag.* **2013**, *34*, 37–46. [Google Scholar] [CrossRef]

35. Stylidis, D.; Biran, A.; Sit, J.; Szivas, E.M. Residents' support for tourism development: The role of residents' place image and perceived tourism impacts. *Tour. Manag.* **2014**, *45*, 260–274. [Google Scholar] [CrossRef]

36. Agenţia de Dezvoltare Regională Nord-Vest. Prezentarea Regiunii Nord-Vest (Transilvania de Nord). Available online: http://www.nord-vest.ro/DESPRE-NOIAgentia-de-Dezvoltare-Regionala-Nord-Vest/REGIUNEA-TRANSILVANIA-DE-NORD/Prezentare-Regiune.html(accessed on 1 November 2015). (In Romanian).

37. The Convention on Biological Diversity. Regiunea Nord-Vest (Cluj-Napoca). Available online: http://biodiversitate.mmediu.ro/romanian-biodiversity/despre-arii-protejate/arpm/regiunea-nord-vest-cluj-napoca/ (accessed on 1 November 2015). (In Romanian).

38. Dumitraş, D.; Pop, A. Perspective on the management of Rodna Mountains National Park.*Bull. UASVM Hortic.* **2009**, *66*, 164–169. [Google Scholar]

39. Dumitras, D.E.; Muresan, I.C.; Ilea, M.; Jitea, I.M. Agritourism—A potential linkage between local communities and parks to maintain sustainability. *Bull. UASVM Hortic.* **2013**, *70*, 300–309. [Google Scholar]

40. EY, Building a Better Working World. Entrepreneurs Speak Out. Entrepreneurship Barometer Romania 2013. Available online: http://www.ey.com/Publication/vwLUAssets/EY_Entrepreneurs_Speak_Out_survey_13_Feb_2014_(EN)/$FILE/EY%20Entrepreneurship%20Barometer%20Romania%202013.pdf (accessed on 30 December 2015).

41. Muresan, I.C.; Arion, F.H.; Harun, R. Study regarding rural guesthouse and tourists' satisfaction. *Bull. UASMV Hortic.* **2013**, *70*, 362–367. [Google Scholar]

42. Ioan, I.; Radulescu, C.V.; Bran, F. Romanian rural tourism: Status and prospects by innovative organizational approaches. *J. Tour. Stud. Res. Tour.* **2014**, *17*, 15–21. [Google Scholar]

43. National Institute of Statistics, Tempo-Online Time Series. Available online: http://statistici.insse.ro/shop/?lang=en (accessed on 25 October 2015).

44. Teye, V.; Sonmez, S.F.; Sirakaya, E. Residents' attitudes toward tourism development. *Ann. Tour. Res.* **2002**, *29*, 668–688. [Google Scholar] [CrossRef]

45. Brida, J.G.; Disenga, M.; Osti, L. Residents' perception and attitudes towards tourism impacts: A case study of the small rural community of Folgaria (Trentino—Italy). *Int. J.* **2011**,*18*, 359–385. [Google Scholar]

46. Abdollahzadeh, G.; Sharifzaden, A. Rural residents' perception toward tourism development: A study from Iran. *Int. J. Tourism Res.* **2014**, *16*, 126–136. [Google Scholar] [CrossRef]

47. Hanafiah, M.H.; Jamaluddin, M.R.; Zulkifly, M.I. Local community attitude and support towards tourism development in Tioman Island, Malaysia. *Procedia Soc. Behav. Sci.* **2013**, *105*, 792–800. [Google Scholar] [CrossRef]

48. Wang, Y.; Pfister, R.E.; Duarte, B. Residents' attitudes toward tourism development: A case study of Washington, NC. In Proceedings of the North-Eastern Recreation Research Symposium, Bolton Landing, NY, USA, 9–11 April 2006; pp. 411–418.

49. Costello, A.B.; Osborne, J.W. Best practices in exploratory factor analysis: Four recommendations for getting the most from your analysis. *Pract. Assess. Res. Eval.* **2005**, *10*, 1–9. [Google Scholar]

50. Jollitte, I.T. *Principal Component Analysis*; Springer Science + Business Media: New York, NY, USA, 2014. [Google Scholar]

51. Ding, C.; He, X. K-means clustering via principal component analysis. In Proceedings of the 21st International Conference on Machine Learning, Banff, AB, Canada, 4–8 July 2004; pp. 29–37.

52. Kaiser, H.F. Index of factorial simplicity. *Psychometrika* **1974**, *39*, 31–36. [Google Scholar] [CrossRef]

53. Tabachinick, B.G.; Fidell, L.S. *Using Multivariate Statistics*, 2nd ed.; Harper & Row: Cambridge, UK, 1989. [Google Scholar]

54. Hair, J.F.; Anderson, R.E.; Tatham, R.L.; Black, W.C. *Multivariate Data Analysis*, 5th ed.; Prentice Hall: Upper Saddle River, NJ, USA, 1998. [Google Scholar]

55. Burgess, S.M.; Steenkamp, J.B.E.M. Marketing renaissance: How to research in emerging markets advances marketing science and practice. *Int. J. Res. Mark.* **2006**, *23*, 337–356. [Google Scholar] [CrossRef]

56. Nunnally, J.; Bernstein, I. *Psychometric Theory*, 3rd ed.; McGraw-Hill: New York, NY, USA, 1994. [Google Scholar]

57. Figueroa-Domecq, C.; Pritchard, A.; Segovia-Perez, M.; Morgan, N.; Villace-Molinero, T. Tourism gender research: A critical accounting. *Ann. Tour. Res.* **2015**, *52*, 87–103. [Google Scholar] [CrossRef]

58. Ministry of Agriculture and Rural Development in Romania. Programul Naţional de Dezvoltare Rurală pentru perioada 2014–2020. Available online: http://www.madr.ro/docs/dezvoltare-rurala/programare-2014-2020/PNDR_2014_-_2020_01.07.2014.pdf (accessed on 22 December 2015). (In Romanian).

59. Ministry of Agriculture and Rural Development in Romania. Situatia proiectelor depuse in cadrul PNDR 2014-2020 la data de 31.12.2015 conform registului electronic al cererilor de finanţare. Available online: http://www.madr.ro/pndr-2014-2020/implementare-pndr-2014-2020/situatia-proiectelor-depuse-2014-2020/download/1816_7595b8fd510ac8d1f2b3cfd18932f357.html (accessed on 19 January 2016). (In Romanian).

4
CHAPTER

Analysis of the Relationship between Tourism and Food Culture

Francisco Javier Jiménez-Beltrán *, Tomás López-Guzmán and Francisco González Santa Cruz

Department of Applied Economics, University of Cordoba, Agrifood Campus of International Excellence ceiA3, Córdoba E-14001, Spai

ABSTRACT

: In recent years, gastronomy has established itself as one of the key elements for the enhancement, sustainable and consolidation of tourist destinations. The aim of this paper is to contribute to the advancement of knowledge on gastronomic tourism in European countries, specifically in the analysis of the relationship between gastronomy, culture and tourism as the research focuses on the city of Córdoba, Spain. The methodology of this research involved conducting surveys with foreign travelers who were lunching or dining at various restaurants in the historic area, and these facilities were characterized by having in their gastronomic menus major typical culinary products of the city using the concept of *tapas*, *i.e.*, the presentation of gastronomy through small portions of food. The results of the study indicate that the healthy component of the gastronomy represents the main dimension. Based on the detected dimensions, three types of international visitors are established (healthy-cultural tourist, cultural tourist and generic tourist) which are considered valid and useful for segmenting the market. This highlights the importance given to gastronomy by tourists as part of the cultural identity of a place and the satisfaction achieved through the gastronomy of the city of Córdoba.

Keywords: gastronomy; tapas; culture; tourism; sustainable

1. INTRODUCTION

Food tourism studies have emerged in recent decades, focusing on food destination, food tourists and hygiene issues, using both qualitative and quantitative analyses [1]. Gastronomic tourists experience a complete sensory experience, especially in terms of flavor, suggesting, according to Cohen and Avieli [2], that through food, tourists receive a greater engagement with the environment where the visit takes place, far from the role of simple observer traditionally associated with tourist visits. In addition, gastronomy is part of the cultural, social, environmental, sustainable and economic history of nations and their people. This is because it reflects a certain lifestyle of the different geographical areas, strengthening traditions in rural areas and modernity in urban areas, as it is something rooted in their own culture and tradition [3], although this implies a constant need for innovation in products and services to provide greater value and thus achieve greater competitiveness in a given location in relation to other destinations [4]. Therefore, local and regional food could give added value to the destination and contribute in this way to the competitiveness of the geographic area [5], and, in this sense, many researchers highlight that each country/region/city should promote food as a central attraction to tourists [6] as dining out and trying national and local cuisines are pleasant activities for most tourists [6,7]. In fact, gastronomy involves a transfer of knowledge and information about the people, culture, traditions and identity of the visited place [8].

Spanish cuisine is known internationally for various aspects such as its innovative nature or the quality of the raw material used. In addition, perhaps one of the most recognizable aspects of this gastronomy is the presentation made of the culinary product through the *tapa* concept which refers to a type of food presentation in small amounts, and it is usually shared among several people even using the same dish. *Tapas* are widely known throughout Spain, although in some regions they are known by another name, *pintxos*, for example, and they allow you to taste a small proportion of the typical gastronomic product of the area along with a drink. In certain geographical areas, the *tapas* that accompany the drink do not even involve an extra cost for the consumer. This small portion of food allows diners to try different types of culinary specialities and thus obtain a wide gastronomic experience. Thus, the typical culinary presentation that takes place in the city of Córdoba, based on the concept of "*tapas*" is one of the most valid options if you want to try many different dishes while sharing food with the rest of diners at a table, achieving in this way greater socialization.

Tapas respond to one of the oldest traditions of Andalusia (and Spain), and today they are established both as heirs of the culinary tradition of this area and as an avant-garde trend by experimenting with new flavors in the cuisine. This allows an exportable model to other countries and, for this reason, in different cities around the world establishments are opening based on the concept of *tapas*.

The aim of this paper is to attempt to answer several questions among which is whether *tapas*can be a tourism resource in a particular city (since it represents a way to experiment with the local cuisine), if *tapas* are a way to better understand the culture of a city or if *tapas* influence the experience of foreign tourists. To this end, this research presents an analysis of the assessments made by foreign travelers regarding different aspects of the concept of *tapas* through a cluster analysis. To achieve this goal, the paper is structured after this introduction in a second paragraph where a literature review is made; a third section that refers to the description of the geographical area; a fourth paragraph where the research methodology is presented; and a fifth paragraph where the results of this research are shown. The paper ends with the conclusions and references used.

2. LITERATURE REVIEW

The academic literature has analyzed gastronomic tourism in different works. Among them, we highlight the work of Hjalager and Richards [9], which brings together a series of articles by leading experts at the time; the work of Boniface [10]; the contributions of Long [11], which includes articles where the identity of gastronomic tourism is addressed; research by Cohen and Avieli [2], the two works coordinated by Professor Hall which analyzes both gastronomy [12] and wine [13] and the contributions of Croce and Perri [14]. In addition, gastronomic tourism is approached by scientific literature in different parts of the world, highlighting especially places located in Asia.This continent is leading the trend in this segment, mainly due to the different conceptions that exist with Western cuisine. Thus, the importance of gastronomic tourism in Hong [15,16], and other Asian countries [2] or in Turkey [17] is analyzed. In addition, in Western countries, various investigations have been conducted focusing on Argentina [18], Canada [19], Chile [20], Croatia [21], Finland [22], Portugal [23] or Spain [24]. On the other hand, studies that have focused on the analysis of a particular type of food are different, as they identify a particular area, although it is known internationally, as it occurs with pizza, sushi or *tapas*. For example, we find religious food [25], Chinese food [26], Indian food [27] or goulash [28].

Henderson presents three research lines around the relationship between tourism and gastronomy [29]: food as a tourism product, the marketing of food to tourists, and food tourism as an instrument of destination and general development. Meanwhile, Tikkanen notes that gastronomy establishes a synergy with tourism through four different aspects: as an attraction, which means that the destination can use this item to promote this place; as a component of the product, where it delves into the design of dining (or oenological) routes; as experience, addressing the existence of one or more locations where gastronomy takes on a different level and becomes a claim in itself, an example of this being the cuisine made by the great masters; and as a cultural phenomenon, premised on the existence of different food festivals [22]. This implies that this type of tourism would have a greater or lesser degree of development according to the level of connection that the cuisine has with the culture of the place where it is settled [30], as the cuisine plays a role in satisfying of the tourist experience and as part of the cultural heritage of the destination [8]. Therefore, this type of tourism reinforces the importance of local ingredients, learning and appreciation of their own consumption, and the importance of culinary resources.

The promotion of a destination through its own gastronomy is carried out by presenting a clear differentiation from its culinary resources, which must have a cuisine that is recognizable by travelers and have a varied and large number of establishments where travelers can enjoy this cuisine. Enright and Newton [31], in their research on Hong Kong, show how the cuisine is the second most important element of attraction, second only to public safety and, as a way of example, even in front of certain tourist attractions. Furthermore, if we consider its relative competitiveness, the cuisine becomes the most important attraction, even surpassing public safety. In short, these authors conclude that the cuisine, public safety, nightlife and visual appeal are the four essential elements for the competitiveness of a destination, if we consider both the relative competitiveness and the importance of the tourist attractions.

As for the segmentation of tourists according to their nationality, distinguishing between domestic and international visitors, there are different studies related to the food and beverage tourism that analyze visitor segmentation in order to perform different exploratory studies. Thus, Alonso *et al.* [32] present the results of an investigation to determine the sociodemographic profiles of wine tourists in New Zealand segmented by nationality. In addition, in the field of food tourism, Nam and Lee [6] present a study on the satisfaction of international visitors in traditional Korean restaurants and Horng *et al.* [33] focus on the analysis of the perception of brand equity in international tourists. On the other hand, there are different studies that focus on the analysis of the perceptions and motivations of foreigners with respect to the gastronomy of a particular place. Thus, we include those of Ghana [34], Croatia [21], Korea [6], Laos [35] and Hong Kong [36].

3. DESCRIPTION OF THE GEOGRAPHICAL AREA

The city of Córdoba is located in southern Spain, in the region of Andalusia. Its population is approximately 325,000 inhabitants and its economy is mainly based on the tertiary sector, emphasizing tourism therein. In 1984, the Mosque-Cathedral of the city was declared a World Heritage Site and subsequently in 1994, so was its Historic Center. Moreover, the Fiesta of the Patios was declared an Intangible Cultural Heritage of Humanity in 2012. Today, the city is one of the major benchmarks of cultural tourism both in Spain and elsewhere in Europe, and it is a meeting point for thousands of travelers who arrive to the city each year attracted by its rich culture, heritage and gastronomy. Focusing on the gastronomy of the city of Córdoba, this is typically based on the Arab cultural heritage, a civilization which flourished for many decades in the city, and whose cuisine is based on both literary and archaeological sources, and it is based primarily on the use of local primary products that are in the geographical area. Within this local raw material, olive oil stands out, because let us recall that the city of Córdoba is located in the geographical area of Andalusia holding the world's largest production of this type of oil. This implies that most of the typical dishes of the city have this product as basis, which is essential in the Mediterranean Diet (declared Intangible Heritage of Humanity in 2010), with multiple functions in each dish: gives flavor, color and aroma; modifies the textures; transmits heat; integrates food; personalizes and identifies a dish and is compatible with all flavors. Let us recall that olive oil is part of the Mediterranean culture itself and, in fact, according to Namdar *et al.* [37], there is evidence of the use of olive oil in Israel in the fifth and sixth millennia BC because they have found traces of olive oil in vessels found in archaeological excavations. With olive oil as a basis, traditional dishes of the Córdoba gastronomy excel such as *salmorejo* (a cold cream made of tomatoes, garlic, bread and olive oil), *flamenquín* (slices of rolled meat in breadcrumbs), and Córdoba Pie (a puff pastry tart made with a pumpkin filling). In addition, olive oil enables tourists visiting this geographic area to also learn more about this cultural heritage through the development of oleotourism. On the other hand, the Córdoba cuisine also differs, as we have indicated above, by presenting its culinary specialities through *tapas* and due to its distinctive culinary establishments. These culinary establishments, called "*tabernas*" (taverns) are catering establishments located in old stately homes and decorated with typical motifs of the Andalusian culture, historically becoming meeting and socialization places of residents and subsequently becoming restaurants. Thus, in these places, there is a synergy between the cultural and historical heritage of the city and the cuisine based on this identity. These taverns are mainly located in the historic area of the city, which enables tourists who are visiting its historical heritage entering into these establishments and learning, through their other senses (which obviously enriches the experience of visitors), of the gastronomic cultural heritage and the new culinary delights of this city.

4. METHODOLOGY

The methodology used in this research was based on conducting fieldwork to determine the valuation of foreign tourists visiting the city of Córdoba on the gastronomy of the city presented through the concept of *tapas*. The surveys were conducted in the establishments selected on two premises: places usually visited by tourists and the selection of dishes and of its cuisine representing the actual gastronomy of the city. The questionnaire was handed to the tourists once they had finished their lunch or dinner. For this reason, talks had been held with the heads of the establishments requesting permission to enter the premises. During the meeting, it was said that the interviewers would seek to avoid any interference with their customer service work.

The survey used in this research is based on various previous works [8,16,17,38,39] and responds to four aspects: socio-demographic characteristics of tourists, satisfaction of the trip, perception and relationship between culture and gastronomy.

Surveys were conducted between the months of April and August 2014. The questionnaire was distributed in four languages (English, French, German and Spanish) using the method of back-translation. Participants completed the survey with complete independence, although the interviewers were present in case they had some kind of difficulty in filling it in. The survey was completely anonymous. Previously, a pre-test of 20 surveys was conducted to detect possible deviations and errors. The total number of questionnaires obtained was 446. A stratification was performed according to the percentages of foreign tourists visiting the city of Córdoba in 2013 and according to data from the NSI (National Statistics Institute) (2015) [40]. Among the tourists surveyed, we can highlight those from France (21.5%), Germany (12.1%), the United Kingdom (12.1%), and the United States (9.2%). A convenience sampling was used, this being commonly used in this type of research where respondents are available to be interviewed in a specific space and time [41].

The items used in the survey are intended to respond to the indicators and measurements proposed for conducting the analysis of the tourist demand [42]. The survey included a 22-item questionnaire, and permits evaluating the motivation or reason of the trip, the socio-economic profile, the knowledge of the typical dishes and the perception of the culinary aspects of the city in order to determine their satisfaction. Thus, a combination of technical issues were used through a Likert 5-point scale to judge the perception, yes/no answers and both open and closed questions, where surveyed tourists were able to make their comments about this gastronomic experience. The total number of tourists who stayed in hotels in the city of Córdoba in 2013 amounted to 364,365 [40], which is considered as the total study population. The research sampling error was 4.64%.Table 1 shows the technical specifications of the research.

Table 1. Technical details of the research.

Number of foreign tourists (2013)	364,365 people
Sample	446 surveys
Sampling error	+/− 4.64%
Performance period	April–August 2014
Procedure	Convenience sampling
Sample Control	Implementation and monitoring of fieldwork by the authors of the research

Source: Own elaboration.

The data collected were organized, tabulated and analyzed using the SPSS 19.0 program (IBM Corporation, Armonk, NY, USA). Data processing was performed through the use of univariate and bivariate statistical tools. Basically, the data, results and conclusions presented in this article refer to the essential attributes that intervene in tasting *tapas* as a culinary resource, the segmentation of the international tourists from these attributes and the satisfaction that they report from the gastronomic point of view. To do this, different statistical techniques were used: the factorial analysis, the cluster analysis, the variance analysis (ANOVA with multiple *post hoc*comparisons) and finally, the chi-square test derived from the contingency tables between variables.

5. RESULTS OF THE INVESTIGATION AND DISCUSSION

The aim of this study is to establish different groups of foreign tourists visiting the city of Córdoba, in their evaluation of several explanatory factors of *tapas* as a way of experiencing the typical gastronomy of the city. These explanatory factors are indicated in Table 2. Foreign tourists surveyed had to assess these explanatory factors using a Likert 5-point scale, 1 being completely disagreeing; 5 being completely agreeing.

Table 2. Explanatory factors regarding the concept of *tapas*.

Explanatory Factor	Denomination
Tapas are an element of socialization	socialization
Tapas allow to taste different dishes prepared with unique culinary arts	tasting
Tapas use fresh culinary ingredients	ingredients
Tapas are easily digestible food	digestion
Tapas are a hygienic form of eating	hygiene
Tapas have a good quality/price ratio	quality/price
The presentation of *tapas* is attractive	presentation
Tapas are a healthy culinary product	healthy
Tapas combine tradition and gastronomic innovation	combination
Tapas are a reflection of the culture of Córdoba	culture

According to these factors, a cluster analysis was performed based on 20 iterations to establish three groups of foreign tourists according to their preferences with respect to *tapas.* Thus, the first cluster has 178 cases, the second 98 and the third 84 cases. The total number of valid cases was 360 while the number of lost cases was 86.

In the first place, a factorial analysis is conducted on the scale of attributes to identify a small number of explanatory factors, obtaining two different dimensions of the attributes referring to*tapas* (Table 3). Although the interest resides in the factorial scores that are derived from these components as a tool to establish the strength of the motivations of each visitor, it is essential to characterize each of the two extracted factors. The first factor is called healthy and represents an attribute related to the appreciation of the local gastronomy due to its greater use of fresh and nutritive ingredients in relation to other types of food. The ingredients used for preparing the *tapas*come from the local agricultural producers, and this implies that the environmental damage is minimized, since there is no transfer of the raw materials from some geographic areas to others. In addition, their use guarantees the maintenance and improvement of the ecological processes. Cronbach's alpha coefficient

(0.86) reveals, for the different items that comprise this motivation dimension, that it represents a reliable subscale. The importance of this factor alone explains 50.27% of the total variance of the motivation matrix. The second factor extracted, called cultural, represents an attribute that finds the more traditional and identitarian meaning of the place where the foods and typical dishes are produced. In addition, it forms part of the social and cultural heritage of the towns, because it reflects a certain style of life of the different geographical areas since gastronomy is something rooted in the culture and local tradition itself [12]. This reinforces the authenticity of the local community, a key aspect for the sustainable development of the region. Cronbach's alpha coefficient (0.79) for this subscale also reveals a good internal consistency. This factor would explain 12.39% of the total variance of the motivation matrix. In addition, the value of the measure of sampling adequacy of Kaiser–Meyer–Olkin (KMO) is = 0.89 and the Bartlett's test of sphericity is 1713.53 with a significance level equal to 0.000. Both results indicate that it is appropriate to perform the factorial analysis. For the extraction of factors, the principal components method was used with Varimax rotation, assigning the items that correspond to each factor.

Table 3. Results of factor analysis and reliability on selection attributes.

Items	Factor Loading		Factor
	1	2	
hygiene	0.81		healthy
healthy	0.80		
digestion	0.78		
ingredients	0.70		
presentation	0.68		
quality/price	0.61		
socialization		0.83	cultural
tasting		0.83	
culture		0.65	
combination		0.58	
Eigen value	5.03	1.24	
% of variance explained	50.27	12.39	
Cronbach's alpha	0.86	0.79	
Kaiser-Meyer-Olkin = 0.89			
Bartlett test of sphericity = 1713.53; $p = 0.000$			
Extraction Method: Principal Component Analysis: Rotation Method: Varimax			

Source: Own elaboration.

The study of the attributes referring to *tapas* provides bases for establishing segmentation of the international tourists that visit the city of Córdoba. For such a purpose, an analysis is made of non-hierarchical clusters with the factorial scores of the two factors analyzed (healthy and cultural). Under the criterion of maximizing the variance between types and minimizing the variance within each type, the best solution that meets the criteria is that of establishing three clusters. The characterization of the clusters from the means of the attribute variables are shown in Table 4. The ANOVA F-statistic shows that the compared means are not equal, but it does not consider where the detected differences are found. In order to know which mean differs from another, a contrast called multiple *post hoc* comparisons or *a posteriori* comparisons were used. With the aim of making these comparisons, it cannot be assumed that the population variances are equal—the critical level associated with the Levene statistic is less than 0.05 in all the cases, rejecting the equality of variances.

Table 4. Level of satisfaction with the visit to the city of Córdoba.

Items	Cluster			ANOVA		Levene Statistic (Significance)	
	1 "cultural" (n = 178)	2 "healthy-cultural" (n = 98)	3 "generic" (n = 84)				
	Mean	Mean	Mean	F	Sig.	Levene	Sig.
hygiene	**3.76(*)**	**4.74(*)**	**2.77(*)**	157.361	<0.01	6.916	<0.01
healthy	**3.66(*)**	**4.80(*)**	**2.54(*)**	233.971	<0.01	22.396	<0.01
digestion	**3.62(*)**	**4.71(*)**	**2.87(*)**	137.011	<0.01	12.045	<0.01
ingredients	**4.06(*)**	**4.87(*)**	**2.83(*)**	231.909	<0.01	13.137	<0.01
presentation	**4.00(*)**	**4.83(*)**	**3.17(*)**	155.185	<0.01	6.752	<0.01
quality/price	**4.27(*)**	**4.73(*)**	**3.27(*)**	117.322	<0.01	9.244	<0.01
socialization	**3.90(*)**	**4.59(*)**	**3.39(*)**	43.240	<0.01	4.141	<0.01
tasting	**4.05(*)**	**4.76(*)**	**3.32(*)**	90.849	<0.01	11.921	<0.01
culture	**4.12(*)**	**4.89(*)**	**3.18(*)**	163.140	<0.01	21.034	<0.01
combination	**3.92(*)**	**4.84(*)**	**2.89(*)**	184.772	<0.01	7.760	<0.01

(*) The values in bold represent significant differences with at least two of the means of the three clusters in ANOVA post-hoc analysis. Source: Own elaboration.

Table 4 shows the evaluation of the explanatory factors for each of the different membership clusters.

In addition, the cases of homoscedasticity and normality have to be fulfilled in the ANOVA F-statistic. Given that it is not possible to assume that the population variances are equal, the Brown–Forsythe and the Welch statistics were used as an alternative to the ANOVA F-statistic (Table S1). Since the critical level associated with both statistics is less than 0.05, the hypothesis of equality of means can be rejected, and it can be concluded that the averages of the variables of the three compared clusters are not equal. For the contrast of the significant differences between the different means, the Games–Howell test was conducted. As for the equality stress tests of means, they are set out in Table S1.

In accordance with Table 4, the first of the clusters presents significant values in some variables, achieving an acceptable score in these items: socialization, tasting, culture and combination. Therefore, it is called cultural tourist, related to the second factor. It is the most numerous cluster with 178 visitors; that is, 49.44% of those surveyed. The second of the clusters presents very significant values in all the attribute variables, achieving a very good score in these items: healthy, ingredients, presentation, culture and combination. In this case, it is called healthy-cultural tourist, related to the first factor as well as with the second. It represents the intermediate cluster in number of visitors (98), which signifies 27.22% of the surveyed. The third of the clusters is comprised of 84 visitors (23.33%), being less numerous and the one with the lowest score of all the variables, especially as regards hygiene, healthy, digestion, ingredients. In this case, it is not related to either of the two factors obtained in the factorial analysis. This cluster can be called generic tourist and its relation to the local gastronomy (represented through *tapas*) is merely biological.

In addition, the concept of sustainability is reinforced. In cluster 2, the element of the cultural heritage (4.89) and the locally produced agricultural products used (4.87) obtain a higher score, delving into the relationship existing between gastronomy, tourism and sustainable development. Furthermore, cluster 1 presents significant results in both items: culture (4.12) and ingredients (4.06), although less conclusive with respect to cluster 2.

Once the three groups of tourists are established and their relationship with the gastronomy of the city of Córdoba through the presentation of *tapas*, three more items are selected: overall rating of *tapas*, satisfied food and gastronomic expectation, all related to the satisfaction of the gastronomic experience during the visit to the city. This is analyzed in Table 5, measured on a Likert 5-point scale, 1 being very dissatisfied and 5 being very satisfied.

Table 5. Level of satisfaction with the visit to the city of Córdoba.

Items	Cluster			ANOVA		Levene Statistic (Significance)	
	1 "cultural" (n = 178)	2 "healthy-cultural" (n = 98)	3 "generic" (n = 84)				
	Mean	Mean	Mean	F	Sig.	Levene	Sig.
Overall rating of tapas	**3.99(*)**	**4.95(*)**	**3.21(*)**	333.989	<0.01	24.447	<0.01
Satisfied food	**4.24(*)**	**4.71(*)**	**3.64(*)**	65.370	<0.01	8.389	<0.01
Gastronomic expectations	**4.11(*)**	**4.68(*)**	**3.40(*)**	84.450	<0.01	13.727	<0.01

(*) The values in bold represent significant differences with at least two of the means of the three clusters in ANOVA post-hoc analysis. Source: Own elaboration.

Having established the tourist groups, it is fundamental to analyze the relations that these clusters may have with the satisfaction of the visit (Table 5). The results show a positive valuation of the gastronomic experience by two of the three segments or tourist groups identified. Overall, the valuations are very high in the second cluster, indicative of the great relevance of the healthy and cultural dimensions, for which reason the gastronomic satisfaction or experience of the visit seems to increase as these dimensions are increased. The significant mean of the second cluster reveals that the cultural dimension alone is also a significant variable, highlighting that the desire to know a culture through its typical gastronomy is also an important source of satisfaction with the destination visited. The third cluster seems to indicate that those visitors who show reasons other than the identified dimensions value to a lesser degree their culinary experience in the city of Córdoba.

The tests of robustness on the means of the satisfaction variable show that the averages of the variables of overall rating of *tapas*, satisfied food and gastronomic expectations in the three compared clusters are not equal. Table S2 shows the equality stress test of means.

Moreover, it is possible to determine the particularities of the identified clusters, by tabulating and crossing them with external variables (socio-demographic data). The variables used are gender, age, education, employment and income (Table 6). The results of Pearson's chi-square test indicate that there are statistically significant differences ($p < 0.05$) between the groups when it deals with the following characteristics: age and employment. With respect to the distribution by sex of those surveyed, clusters 1 and 2 contain a greater male representation compared to the female, while, in cluster 3, the proportion is inverted. As for age, cluster 1 contains the higher percentage of tourists situated between 30 and 49 years of age (54.80%), while cluster 3 presents the highest values in the extreme segments: from 16 to 30 years (28.90%) and 60 years and older (16.90%).

Table 6. Socio-demographic characteristics of three clusters of tourists visiting Córdoba.

Socio-Demographic Characteristics	1 "cultural" (n = 178)	2 "healthy-cultural" (n = 98)	3 "generic" (n = 84)	Total (%)	Statistics
Gender	100.00	100.00	100.00	100.00	Chi2 = 5.569; $p = 0.473$
Male	53.67	56.12	43.37	52.00	
Female	46.30	43.90	56.60	48.00	
Age	100.00	100.00	100.00	100.00	Chi2 = 20.041; $p = 0.010$
Under 30 years	22.60	26.80	28.90	25.20	
30–39 years old	30.50	19.60	20.50	25.20	
40–49 years old	24.30	19.60	9.60	19.60	
50–59 years old	14.10	24.70	24.10	19.30	
60 years old or more	8.50	9.30	16.90	10.60	
Education	100.00	100.00	100.00	100.00	Chi2 = 5.646; $p = 0.227$
Primary education	3.40	1.00	1.20	2.20	
Secondary education	17.10	26.80	19.00	20.20	
University education	79.50	72.20	79.80	77.60	
Employment	100.00	100.00	100.00	100.00	Chi2 = 21.025; $p = 0.021$
Employee	14.60	25.00	113.40	17.10	
Civil Servant	52.60	43.50	35.40	46.10	
Self-employed	15.20	13.00	14.60	14.50	
Student	7.00	8.70	18.30	10.10	
Retired	8.80	9.80	17.10	11.00	
Income	100.00	100.00	100.00	100.00	Chi2 = 7.466; $p = 0.487$
Under €700	11.70	12.00	17.60	13.10	
From €700 to €1000	9.40	4.30	5.40	7.10	
From €1001 to €1500	14.00	14.10	16.20	14.50	
From €1501 to €2500	33.30	29.30	23.00	30.00	
Over €2500	31.60	40.20	37.80	35.30	

Source: Own elaboration.

On the other hand, the educational level of those surveyed particularly stands out: in cluster 2, 72.20% state they have a university degree, while those that have a secondary education are 26.80%. In the professional category, it is important to point out that, in cluster 2, the figures of civil servant (43.50%) and employee (25.00%) are predominant. In the economic section, for the three clusters, there is an elevated concentration (over 60.00%) in the two higher intervals defined for the salary level (more than €1501). Specifically, cluster 2 presents the highest results—over €2500 (40.20%). In short, the results are a reliable reflection of the high purchasing power that characterizes the foreign tourist demand that visits Córdoba, providing a unique opportunity to generate income and employment for the region.

6. CONCLUSIONS

Tourists visiting a particular cultural destination, in addition to participating in the knowledge of their heritage, want to increase their sensory experience mainly through gastronomy. In this sense, gastronomy and its relation to tourism has become a key aspect in the analysis of tourist destinations, especially those related to culture and heritage.

The city of Córdoba has traditionally been an important cultural destination due to the importance of its historical heritage. In order to complete the tourist experience, in recent years, the typical regional gastronomy has been reinforced, offering tourists the typical culinary dishes of the city. However, in addition, these specialities are being introduced through a concept that defines the Spanish cuisine in general and that of the region of Andalusia especially—*tapas*.

The heterogeneity in the preferences of the consumers make it necessary to go into depth in the knowledge of the reasons that may impact the processes of choice in order to try the local gastronomy. The analysis made has identified two different factors or attributes: healthy and cultural. Based on them, three types of visitors were established—cultural tourist, healthy-cultural tourist and generic tourist that are considered as valid and of use for segmenting the market. The cultural tourist is the most numerous (49.44%) and gets the highest scores on the items related to the cultural dimension (tasting and culture). Next, the healthy-cultural tourist is the most important from the point of view of this research. It groups the second collective of travelers (27.22%) and presents the highest values in all the items, both in the healthy dimension as in that of culture. It closely reflects the use of fresh culinary ingredients and how *tapas* reflect part of the culture of the city of Córdoba.

The results obtained, as regards the degree of satisfaction with the gastronomy are, in general, high. However, there are significant differences between the identified tourist segments, being clearly linked to the analyzed dimensions. The highest degree of satisfaction, by far, corresponds to the tourists interested in the healthy and cultural aspects of the local gastronomy through *tapas*. The satisfaction goes down slightly, although it remains high in those tourists who are concentrated on the cultural aspect.

The main results of this research, conducted through a cluster analysis indicate that the main characteristics that define *tapas* are highly valued, in general, by all the groups analyzed, thus reinforcing the importance of the gastronomy of the place, and its presentation, as part of the experience in a particular destination.

This paper contributes to the concept of sustainability, as one of its main conclusions is that the potentiation of local gastronomy as a tourist product involves strengthening local economies, a close relationship between local agriculture and tourism, clearly enhancing the environment in two different ways: increasing the level of both organic production and local distribution. In the same way, it is necessary to guarantee respect for tradition and the values that make up the authenticity of the host community. In fact, the use of the local and seasonal agricultural products (more respectful of the environment) and the cultural heritage associated with the concept of *tapas* of the city of Cordoba are considered some very valued elements. Definitely, market segmentation is the way through which these preferences (with respect to *tapas*) of tourists can be included to allow the development of an appropriate marketing strategy and to ensure the continuous growth and sustainability of the local gastronomy.

The main limitations of this research are the time period in which it is conducted, which makes us believe that it would be desirable to extend the investigation to tourism arriving to the city during all months of the year.

As the main future line of research, we recommend strengthening research to see if tourists visiting the city of Córdoba and consuming its gastronomy also consume food products of this area later in their home towns.

ACKNOWLEDGMENTS

The authors thank typical taverns and restaurant of Cordoba that have participated in this research as well as the city council of Córdoba.

AUTHOR CONTRIBUTIONS

The research method was proposed and conducted by Francisco Javier Jiménez-Beltrán. The structure of the research project was discussed between Francisco González Santa Cruz, Tomás López-Guzmán and Francisco Javier Jiménez-Beltrán. The paper has been written by Francisco Javier Jiménez-Beltrán and Tomás López-Guzmán.

CONFLICTS OF INTEREST

The authors declare no conflict of interest.

REFERENCES

1. Lee, K.H.; Scott, N. Food tourism reviewed using the paradigm funnel approach. *J. Culin. Sci. Technol.* **2015**, *13*, 95–115. [Google Scholar] [CrossRef]

2. Cohen, E.; Avieli, N. Food in tourism: Attraction and Impediment. *Ann. Tour. Res.* **2004**, *31*, 755–778. [Google Scholar] [CrossRef]

3. Mitchell, R.; Hall, C.M. Wine tourism research: the state of play. *Tour. Rev. Int.* **2006**, *9*, 307–332. [Google Scholar] [CrossRef]

4. Richards, G. Gastronomy: An essential ingredient in tourism production and consumption? In *Tourism and Gastronomy*; Hjalager, A.M., Richards, G., Eds.; Routledge: London, UK, 2002; pp. 3–20. [Google Scholar]

5. Crouch, G.I.; Richie, J.R.B. Tourism, competitiveness, and societal prosperity. *J. Bus. Res.* **1999**, *44*, 137–152. [Google Scholar] [CrossRef]

6. Nam, J.H.; Lee, T.J. Foreign travelers' satisfaction with traditional Korean restaurants. *Int. J. Hosp. Manag.* **2011**, *30*, 982–989. [Google Scholar] [CrossRef]

7. Harrington, R.J.; Ottenbacher, M.C. Culinary tourism: A case study of the gastronomic capital. *J. Culin. Sci. Technol.* **2010**, *8*, 14–32. [Google Scholar] [CrossRef]

8. Ignatov, E.; Smith, S. Segmenting Canadian Culinary Tourists. *Curr. Issues Tour.* **2006**, *9*, 235–255. [Google Scholar] [CrossRef]

9. Hjalager, A.M.; Richards, G. *Tourism and Gastronomy*; Routledge: London, UK, 2002. [Google Scholar]

10. Boniface, P. *Tasting Tourism: Travelling for Food and Drin*; Ashgate Publishing Limited: Aldershot, UK, 2003. [Google Scholar]

11. Long, L.M. Culinary tourism: A folkloristic on eating and otherness. In *Culinary Tourism*; Long, L.M., Ed.; The University Press of Kentucky: Lexington, KY, USA, 2004; pp. 20–50. [Google Scholar]

12. Hall, M.C.; Sharples, L.; Mitchell, R.; Mcionis, N.; Cambourne, B. *Food Tourism around the World*; Butterworth-Heinemann: Oxford, UK, 2003. [Google Scholar]

13. Hall, C.M.; Sharples, L.; Cambourne, B.; Macionis, N. *Wine Tourism around the World: Development, Management and Markets*; Elsevier: Oxford, UK, 2000. [Google Scholar]

14. Croce, E.; Perri, G. *Food and Wine Tourism*; Cabi: Oxford, UK, 2010. [Google Scholar]

15. Au, N.; Law, R. Categorical classification of tourism dining. *Ann. Tour. Res.* **2002**, *29*, 819–833. [Google Scholar] [CrossRef]

16. McKercher, B.; Okumus, F.; Okumus, B. Food Tourism as a Viable Market Segment: It's All How You Cook the Numbers! *J. Travel Tour. Mark.* **2008**, *25*, 137–148. [Google Scholar] [CrossRef]

17. Okumus, B.; Okumus, F.; Mckercher, B. Incorporating local and international cuisines in the marketing of tourism destinations: The cases of Hong Kong and Turkey. *Tour. Manag.* **2007**,*28*, 253–261. [Google Scholar] [CrossRef]

18. Schülter, R.; Gándara, J.M. *Gastronomía y Turismo. Una Perspectiva*; CIET: Buenos Aires, Argentina, 2003. [Google Scholar]

19. Stewart, J.W.; Bramble, L.; Ziraldo, D. Key challenges in wine and culinary tourism with practical recommendations. *Int. J. Contemp. Hospit. Manag.* **2008**, *20*, 303–312. [Google Scholar]

20. Ascanio, A. Rutas gastronómicas chilenas: Una aproximación al tema. *Pasos Rev. Tur. Patrim. Cult.* **2009**, *7*, 321–325. (In Spanish) [Google Scholar]

21. Fox, R. Reinventing the gastronomic identity of Croatian tourist destinations. *Int. J. Hosp. Manag.* **2007**, *26*, 546–559. [Google Scholar] [CrossRef]

22. Tikkanen, I. Maslow's hierarchy and food tourism in Finland: Five cases. *Br. Food. J.* **2007**,*109*, 721–734. [Google Scholar] [CrossRef]

23. Oliveira, S. La importancia de la gastronomía en el turismo. Un ejemplo de Mealhada. *Estud. Persp. Tur.* **2007**, *16*, 261–282. (In Spanish) [Google Scholar]

24. López-Guzmán, T.; Di-Clemente, E.; Hernández-Mogollón, J.M. Culinary tourists in the Spanish region of Extremadura,

Spain. *Wine Econ. Policy* **2014**, *3*, 10–18. [Google Scholar] [CrossRef]

25. Son, A.; Xu, H. Religious food as a tourism attraction: The roles of Buddhist temple food in Western tourist experience. *J. Herit. Tour.* **2013**, *8*, 248–258. [Google Scholar] [CrossRef]

26. Cheung, S.C.M. From foodways to intangible heritage: A case study of Chinese culinary resource, retail and recipe in Hong Kong. *Int. J. Herit. Stud.* **2013**, *19*, 353–364. [Google Scholar] [CrossRef]

27. Chhabra, D.; Lee, W.; Zhao, S.; Scout, K. Marketing of ethnic food experiences: Authentication analysis of Indian cuisine abroad. *J. Herit. Tour.* **2013**, *8*, 145–157. [Google Scholar] [CrossRef]

28. Metro-Roland, M.M. Goulash nationalism: The culinary identity of a nation. *J. Herit. Tour.* **2013**, *8*, 172–181. [Google Scholar] [CrossRef]

29. Henderson, J.C. Food tourism reviewed. *Br. Food. J.* **2009**, *111*, 317–326. [Google Scholar] [CrossRef]

30. Riley, M. Food and beverage management: A review of change. *Int. J. Contemp. Hospit. Manag.* **2005**, *17*, 88–93. [Google Scholar] [CrossRef]

31. Enright, M.J.; Newton, J. Tourism destination competitiveness: A quantitative approach. *Tour. Manag.* **2004**, *25*, 777–788. [Google Scholar] [CrossRef]

32. Alonso, A.D.; Fraser, R.A.; Cohen, D.A. Investigating differences between domestic and international winery visitors in New Zealand. *Int. J. Wine Bus. Res.* **2007**, *19*, 114–126. [Google Scholar] [CrossRef]

33. Horng, J.S.; Liu, C.H.; Chou, H.Y.; Tsai, C.Y. Understanding the impact of culinary brand equity and destination familiarity on travel intentions. *Tour. Manag.* **2012**, *33*, 815–824. [Google Scholar] [CrossRef]

34. Amuquandoh, F.E.; Asafo-Adjei, R. Traditional food preferences of tourism in Ghana. *Br. Food. J.* **2013**, *115*, 987–1002. [Google Scholar] [CrossRef]

35. Staiff, R.; Bushell, R. The rhetoric of Lao/French fusion: Beyond the representation of the Western tourist experience of cuisine in the World Heritage City of LuangPrabang, Laos. *J. Herit. Tour.* **2013**, *8*, 133–144. [Google Scholar] [CrossRef]

36. Tse, P.; Crotts, J.C. Antecedents of novelty seeking: International visitors' propensity to experiment across Hong Kong's culinary traditions. *Tour. Manag.* **2005**, *26*, 965–968. [Google Scholar] [CrossRef]

37. Namdar, D.; Amrani, A.; Getzov, N.; Milevski, I. Olive oil storage during the fifth and sixth millennia BC at EinZippori, Northern Israel. *Isr. J. Plant. Sci.* **2014**, *62*, 65–74. [Google Scholar] [CrossRef]

38. Dawson, H.; Holmes, M.; Jacobs, H.; Wade, R.I. Wine tourism: Winery visitation in the wine appellations of Ontario. *J. Vacat. Mark.* **2011**, *17*, 237–246. [Google Scholar] [CrossRef]

39. López-Guzmán, T.; Sánchez-Cañizares, S. Culinary tourism in Cordoba (Spain). *Br. Food. J.* **2012**, *114*, 168–179. [Google Scholar] [CrossRef]

40. National Statistics Institute. *Hotel Occupancy Survey*; INE Publications: Madrid, Spain, 2015. [Google Scholar]

41. Finn, M.; Elliott-White, M.; Walton, M. *Tourism and Leisure Research Methods: Data Collection, Analysis and Interpretation*; Pearson Education: Harlow, UK, 2000. [Google Scholar]

42. Getz, D.; Brown, G. Critical success factors for wine tourism regions: A demand analysis. *Tour. Manag.* **2006**, *27*, 146–158.

5
CHAPTER

Poverty Alleviation through Pro-Poor Tourism: The Role of Botswana Forest Reserves

Haretsebe Manwa [1,†,*] and Farai Manwa [2,†]

[1] *Department of Tourism, North West University, Private Bag X2046, Mmabatho, Mafikeng 2735, South Africa*

[2] *International Trade, Southern Cross University, Tweed Heads NSW 2485, Australia*

ABSTRACT

Both government and international donor agencies now promote the use of tourism to alleviate poverty. The Botswana government has embraced tourism as a meaningful and sustainable economic activity and diversification opportunity, which now ranks second after mining in its contribution to the country's gross domestic product. The study reported in this paper investigates perceptions of stakeholders on the opportunities that would be created for the poor by opening up Botswana's forest reserves for ecotourism. Data was collected through mixed methods involving in-depth interviews with government departments, traditional leaders, quasi-government organisations and the Hospitality and Tourism Association of Botswana. Focus group discussions were also held with village development committees, Chobe Enclave Conservation Trust (CECT) and Kasane, Lesoma and Pandematenga Trust (KALEPA) members, and a consultative national workshop of stakeholders was also held. The findings indicate that opening up forest reserves for ecotourism has the potential to alleviate poverty among the disadvantaged groups living adjacent to forest reserves through direct (employment, small- and medium-sized enterprises (SMEs)), secondary (linkages/partnerships) and dynamic effects (sustainable livelihoods). The study concludes by cautioning that whilst pro-poor tourism may yield short- and medium-term benefits, in keeping with sustainability objectives, participants in the programme need to be mindful of forestry encroachment and come up with strategies to ensure the sustainability of the Botswana forest reserves.

Keywords: pro-poor tourism; forest reserves; sustainable tourism; institutional capacity building; Botswana

1. INTRODUCTION

Botswana has achieved the status of a middle-income country as a result of its mineral wealth, mainly through the exploitation of its diamond reserves. Despite this economic leap, not everybody has benefitted. It has been reported that, countrywide, 20.7% of the population live below the poverty datum line [1], and poverty has been particularly persistent in the most remote areas of northern and western Botswana [2]. The volatility of mineral resources and their non-renewability has prompted Botswana to look for alternative and sustainable economic activities. Tourism has been perceived as a meaningful diversification of the economy. More importantly, tourism is professed as being able to alleviate poverty among the poor, the majority of whom live in tourism resource-rich areas [3,4,5]. Therefore, tourism becomes an ideal candidate for poverty alleviation in Botswana, due to its labour-intensive nature, which enables it to employ disadvantaged members of the community with very little formal training, e.g., women and youth. In addition, tourism utilizes natural and cultural resources of the poor, whereby they can utilize these resources for their benefit [6].

However, tourism is a private sector business and, therefore, is controlled by market forces. By its very nature, tourism is geared to serve the interests of the tourists and investors and, therefore, does not automatically improve the welfare of the poorer members of the community [7]. In addition, tourism can be viewed as a form of development strategy whose agenda is driven by multinational corporations promoting neo-colonial states in the least developed countries [8], and thus, caution should be exercised when advocating for tourism as a one-size fits all strategy for poverty alleviation.

This study aims to build on the extant literature on pro-poor tourism development. This is achieved through an investigation of how the Botswana Forest Reserves can be instrumental in alleviating poverty among local communities through the use of pro-poor tourism strategies.

The study uses Mitchell and Ashley's [9] framework, to analyse the potential opportunities that would be created for the poor through the opening of Botswana's forest reserves for ecotourism. The paper draws its data from the views of the Botswana Forest Reserves' stakeholders. In particular, the paper is guided by the following research questions:

- What attractions prevalent in the forest reserves could be used for ecotourism development?
- What are the likely costs and benefits to the communities from ecotourism development in the Botswana Forest Reserves?
- Are there differences in the roles that can be played by the private sector vis-à-vis the roles of the local communities in ecotourism development in the reserves?

The Botswana Forest Reserves were chosen due to the bulk of tourism research in Botswana having been concentrated in the Okavango Delta [10,11,12] with little effort directed towards other tourism facets, such as cultural tourism [13] and forest-based tourism [14]. The paper provides a brief survey on the literature on pro-poor tourism, followed by background information on the case study area. The next section outlines the main methodological approaches and then presents the results and discussion section. The study concludes by cautioning that, whilst pro-poor tourism may yield short-and medium-term benefits, in keeping with sustainability objectives, participants in the programme need to be mindful of forestry encroachment and come up with strategies to ensure the sustainability of the Botswana Forest Reserves.

2. THEORETICAL FRAMEWORK

The impact of tourism in poverty alleviation is now a growing area of research. This is demonstrated through case studies from many different parts of the world [6,15,16,17,18,19,20,21]. Its roots are embedded in the sustainable tourism framework, which, in turn, is derived from the overarching field of sustainable development. According to Tribe [22] (p. 298), sustainability can be defined as growth that does not encourage resource depletion or social unrest. In the tourism context, this means a level of development that is in equilibrium with the carrying capacity of the destination and that does not alter the current ecosystem, whilst not disadvantaging future generations through resource depletion [23].

Sustainability has tended to focus on broad ecological maintainable principles, policy objectives and management techniques employed in destination management. There has been significantly less academic literature on the equitable distribution of natural resources and the accrual of externalities from the resources in question [24] (p. 864). Pro-poor tourism has been put forward as a means of addressing this gap through fostering development that is inclusive of under-privileged members of society, with the proceeds from its activities going towards improving the welfare of the community members [25] (p. 208).

Pro-poor tourism is now advocated as a means of alleviating poverty in developing economies [26,27,28,29,30,31]. The pro-poor framework has generated a lot of interest and promotion, not only among researchers, but also amongst international and donor organisations. For example, the United Nations World Tourism Organisation has come up with the Sustainable Tourism Eliminating Poverty Initiative (STEP), which began in 2002. The focus of STEP is to encourage the promotion of activities that promote sustainable tourism (social, economic, ecological) with a focus on alleviating poverty through the development and creation of jobs for people living on less than a dollar a day [32].

The Department for International Development (DFID) in the United Kingdom [33] (p. 1) has defined pro-poor tourism as "tourism that generates net benefits for the poor" (*i.e.*, benefits are greater than costs). DFID further clarifies that benefits should be viewed as holistic and should not be restricted to economic benefits, but should also include social, environmental and cultural benefits. In addition, pro-poor tourism should not be viewed as a tourism product, but rather as an approach to tourism development and management through which linkages are developed between tourism businesses and poor people as a way of leveraging and increasing the tourism benefits to the poor [34] (p. 2001).

Mitchell and Ashley [9] have extended the understanding of how to determine the impacts of tourism on poor communities through the development of a framework that uses three pathways through which the benefits of tourism can be transferred to the poor. They categorise these into direct effects, secondary effects and dynamic effects [35] (p. 3) (see Figure 1). Direct effects are realised through direct employment in the tourism industry. Examples of this include community employment by corporations [36], leasing of communal land to tourism operators, the establishment of business enterprises, such as curio shops or craft markets by community members, and cultural tourism [6,37,38]. Whilst many authors note that direct effects can increase economic benefit to the community in a variety of ways [39,40,41,42,43,44], Mitchell and Ashley [9] caution that externalities are not always positive, with losses of livelihood to the community occurring in some instances.

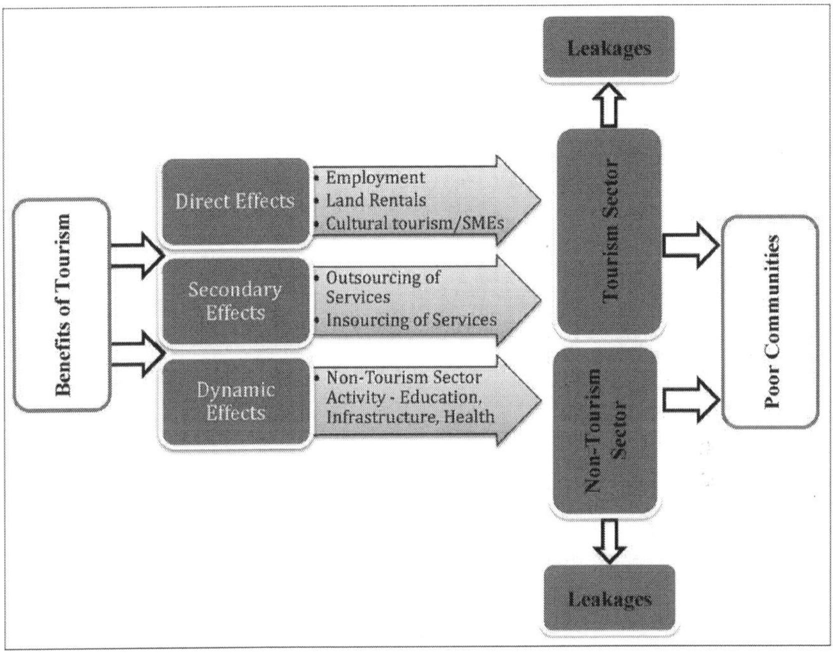

Figure 1. Pathways of benefits to the poor adapted from Mitchell and Ashley, 2010. SMEs, small- and medium-sized enterprises.

For secondary effects to occur, there needs to be a shift in the private sector mind-set and a willingness to collaborate with local communities through the establishment of linkages [45,46,47,48,49,50]. Examples of such linkages include tourism operators sourcing food and beverage supplies from local communities [20,46]. Other possible areas include "in-sourcing", which refers to a corporation hiving off an operation; for example, the provision of opportunity for staff employed in housekeeping to run their unit as a business, eventually leading to the removal of these employees from the company's payroll [47,48].

In the third pathway, the dynamic effects are structural changes, which are indirectly attributed to tourism development. For example, tourism development can result in self-sufficiency, where communities have in-house facilities developed, such as grinding mills, tap water, schools and clinics [12,49,50].

On the other hand, pro-poor tourism has received a fair amount of criticism [51]. It has been argued, for example, that pro-poor tourism perpetuates inequalities prevailing in society, since both rich and poor benefit from pro-poor tourism [7]. Holden *et al.* [18] (p. 331) argue that there are inherent structural barriers militating against the poor involved and benefitting from tourism. They highlight such barriers as low-level education attainment, lack of micro-finance targeted at tourism development and marginalisation from decision-making. Another interesting point raised by Scheyvens [7] (p. 91) is the misconception that "tourism industry operators should have some ethical commitment to ensuring that their businesses contribute to local poverty-alleviation". Schilcher [8] as a result argues for protectionist measures to ensure that big multinational players do not set the agenda and monopolise decision-making in the global arena.

3. METHODS

3.1. Study Area

This paper is based on a study that was carried out in the Chobe District, which is one of Botswana's leading tourism destinations in addition to the Okavango Delta. The Chobe District is a small region covering an area of 22,052 km^2 and a total population of 23,449 people [2]. About 50% of the district is taken up by the Chobe National Park, and a further 4096 km^2 is taken up by forest reserves. The bulk of the land is reserved for conservation, and about 31% is left for communal use (residential, arable, livestock, grazing and settlement expansion) [52]. The shortage of land in the Chobe District is an impediment to the growth of the agricultural sector, with the bulk of arable land reserved mainly for conservation [52].

Wildlife viewing and safari hunting dominate the tourism activities that are undertaken in the Chobe District [53]. Consequently, this activity has resulted in congested development in the form of lodges and hotels, all competing for the river and wildlife resources. Using forest reserves for ecotourism would open up alternative tourism development areas, which would alleviate the pressure on the Chobe riverfront. The study therefore explored tourism in forest reserves as a sustainable tourism option for communities to benefit from using the Mitchell and Ashley Framework as the analytical tool. The study focused on four villages located in Chobe West (Mabele, Kachikau, Kavimba and Satau) and three villages in Chobe East (Kasane, Lesoma and Pandamatenga), which were all adjacent to the forest reserves (see Figure 2).

The Botswana Forest Reserves comprise six gazetted protected areas: Chobe Forest Reserve, Kasane Extension Forest Reserve, Kasane Forest Reserve, Kazuma Forest Reserve, Maikaelelo Forest Reserve and Sibuyu

Forest Reserve (Table 1). Forest reserves occupy around 0.8% of the country [54]; however, this translates into 22% of the Chobe District land (Chobe District Settlement Strategy 2005). The forest reserves (FRs) constitute an area of high biological diversity in terms of both flora and fauna. They are important areas for wildlife migration within the region and form part of the Miombo Woodlands, which extend to parts of other Southern African Development Community countries, such as Angola, Malawi, Zambia and Zimbabwe. They are also part of the Kavango-Zambezi (KAZA) Transfrontier conservation area in the Kavango and Zambezi river basins (stretching into Angola, Namibia, Zambia and Zimbabwe).

The forest reserves were gazetted with the purpose of conserving and regulating the use of the unique forests of northern Botswana. Initially, forest reserves were protected to safeguard and control commercial logging of hardwoods *Baikiaea plurijuga* and *Pterocarpus angolensis* for the export market [54]. Failure to supervise operations and resource wastage of timber resources led to the cessation of commercial logging in 1992 [14].

Table 1. Forest reserves in Botswana.

Name	Since	Size in km^2
Kasane Forest Reserve	1968	109.9
Kasane ext. Forest Reserve	1981	641.1
Chobe Forest Reserve	1981	1485
Kazuma Forest Reserve	1981	156
Maikaelelo Forest Reserve	1981	543
Sibuyu Forest Reserve	1981	1161
Totals		4096

Source: Norwegian Forestry Society, 1993.

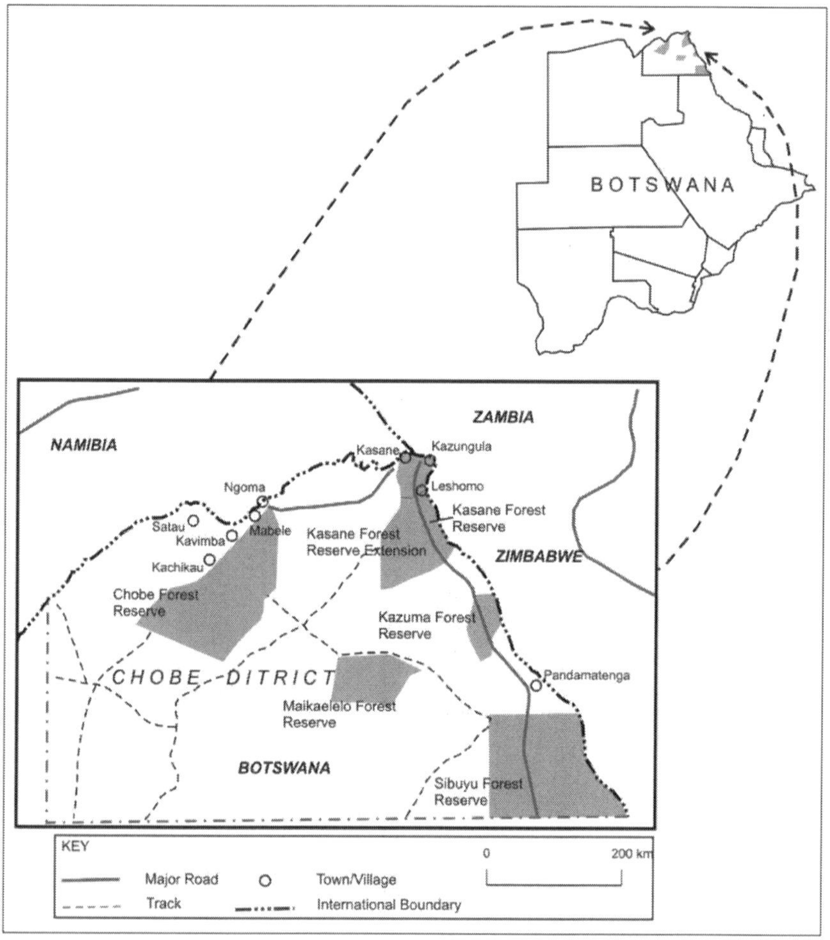

Figure 2. Map of the location of the Botswana Forest Reserves.

3.2. Data Collection

Data were collected through triangulation or mixed methods [55] using a combination of secondary and primary data sources. Secondary data constituted existing literature, government reports and official documents, book chapters and other relevant documentation. Primary data was derived by means of thematic interviews targeted at senior officers in central government in Gaborone and district offices, as well as focus group discussions with community groups.

A total of 18 people were interviewed. The interviewees represented 6 government departments, 4 quasi-government organisations, 3 chiefs, 3 tourism private sector organisations and 2 non-governmental organisations that deal with conservation and human rights. The interviews, whose duration ranged between 30 and 40 min, were held between February and March, 2009. The interview guidelines addressed the study questions as follows:

Theme 1: What attractions prevalent in the forest reserves could be used for ecotourism development? The questions covered tourism activities taking place in the reserves, natural and cultural resources in the forest reserves and what tourism activities and facilities should be developed in the forest reserves. Also included were the perceptions on the forest reserves' potential for ecotourism development.

Theme 2: What are the likely costs and benefits to the communities of opening the forest reserves for ecotourism development? Among the areas covered were potential conflict, possible costs and benefits to local communities of opening up the reserves for ecotourism. Opinions were sought on how conflicts and the cost of conflicts could be minimized? Other areas of interest included community empowerment through ecotourism development and community attitudes towards forest reserves and their use.

Theme 3: Are there differences in the roles that can be played by the private sector vis-à-vis the roles of the local communities in ecotourism development in the Botswana Forest Reserves? The interviews assessed perceptions on the types of ecotourism projects that could be developed by communities and the private sector. Questions relating to location were posed, as well as questions relating to the diversification and improvement of local livelihoods.

Focus group discussions were also held in May, 2009, with the Local Enterprise Authority (LEA), which is an organization that is responsible for the development of small- and medium-sized enterprises (SMEs) in Botswana. LEA pursues this mandate by offering financial and human resources training and internships to SMEs, community leaders, women's groups, village development committees and trust members of the four villages in Chobe West (Mabele, Kachikau, Kavimba and Satau) and the three villages in Chobe East (Kasane, Lesoma and Pandamatenga). We opted for village development committees instead of households, because we believed that such committees would be able to more appropriately articulate the concerns of their members than normal households. Moreover, village development committees play a pivotal role in catering for local needs. The focus groups pursued similar issues to those covered under the individual interviews.

The final results from both of the interviews and focus group discussions were presented to representatives of the stakeholders at a national workshop held in May, 2009, aimed at validating the findings. Emerging themes analysis was adopted as the primary analysis tool with which to draw out data from the responses. The responses were analysed in relation to the themes presented and their implications to the development of pro-poor tourism in Botswana's forest reserves.

4. RESULTS

The aim of the study was to investigate stakeholder perceptions on how the Botswana Forest Reserves can be instrumental in alleviating poverty among local communities through the use of pro-poor tourism strategies. The perceptions of the respondents of the study were presented according to the research questions, which guided the study as reflected in Table 2. The results were then further refined, as shown in Table 3, in terms of the perceived effects using the Mitchell and Ashley framework.

Table 2. Pro-poor tourism themes.

Resources	Potential ecotourism products
1. Wildlife	Photographic safaris, game drives, walking safaris, bird watching, night drives, walking trails, night drives
2. Infrastructure	Old boreholes, watering holes, wells, camps, lodges, temporary structures
3. Cultural and Heritage Tourism	Graves, artefacts and tools used by the ancestors, historical sites (remnants of Basarwa settlements), dance groups, food, way of life of surrounding villages
4. Agriculture	Crop farming, availability of wild fruits and thatched grass
5. Costs	Loss of traditional livelihoods, depletion of forest resources (e.g., medicinal plants, wild fruits), desecration of graves, exploitation and leakages of wealth
6. Benefits	Development of facilities, improved livelihoods, employment opportunities, development of community-based tourism businesses, development of small- and medium-sized enterprises, partnerships with the private sector to run tourism businesses
7. Community Involvement	Contribute as part of the workforce, formation of community-based tourism companies (CBT), formation of partnerships with the private sector
8. Private Sector Involvement	Provision of tour companies (safari companies), marketing of destination (including community-based), training and empowerment (Human capital), development of packages including CBTs

4.1. Perceptions on Tourism Attractions and Ecotourism Development

The perceptions from the study were that the Botswana Forest Reserves are rich in wildlife (including birds). Ecotourism activities that could be undertaken in the forest reserves included photographic safaris, bird watching, game drives, walking safaris and walking trails. Cultural and heritage tourism was another ecotourism attraction emphasised by the stakeholders. The forest reserves are rich in cultural resources, notably graves, artefacts, old tools used by Basarwa (Bushmen) and the remnants of old settlements. In addition, the respondents highlighted that the local communities have a unique culture in terms of food, dance and arts, which could be incorporated into the cultural experience of tourists to the forest reserves. This would ensure that tourists stayed longer and, as a result, spend more money at the destination.

Table 3. Forest reserve activity classification. KALEPA, Kasane, Lesoma and Pandematenga Trust.

	Direct Effects			
	Tourism Activity	**Tourism Sector**	**Leakage**	**Perception**
1.	Game drives, photographic safari	Wildlife Viewing	Yes	Employment with multiplier effect on the community; transfer of business skills, but not in all circumstances; community-run; direct impacts on livelihoods; human capital development
2.	Privately-run lodge development and tented camps	Accommodation/ Heritage Tourism	Yes	Supply labour and outsource some functions of the lodge (e.g., agricultural produce)
3.	Campsites to be run by Ngoma, Mabele, Kavimba and Kachikau communities	Accommodation	No	Job creation and empowerment
4.	Community-run accommodation facilities (camps, lodges)	Accommodation	No	Use of local materials, capacity building and empowerment
5.	Mobile safaris/campsite through public private partnership with Lesoma community	Accommodation/ Wildlife Viewing	Yes	Job creation and empowerment

Table 3. *Cont.*

	Direct Effects			
	Tourism Activity	**Tourism Sector**	**Leakage**	**Perception**
6.	Campsites, private public partnership with community-run accommodation with a heritage site to provide interpretation, community-run campsite (KALEPA)	Accommodation/Heritage Tourism	No	Job creation through outsourcing to the community employment opportunities in heritage tourism activities; use of resources from the community, with the possibility of exploitation
7.	SMEs (crafts, artefacts, souvenirs)	SMEs	No	Quality issues; lack of business skills, e.g., bookkeeping and marketing
	Secondary Effects			
	Tourism Activity	**Tourism Sector**	**Leakage**	**Perception**
1.	Sourcing of food supply from local community	Hotels/Restaurants	No	Promote self-sufficiency and an increase in food production; sustainable livelihoods
2.	Village tours, selling of artefacts, souvenirs, local cuisine, dance troupes/groups	Cultural and Heritage Tourism	Yes	Promote cultural awareness and job creation; fair trade; lacking in administrative skills and access to tourists
	Dynamic Effects			
	Community Related Activity	**Sector**	**Leakage**	**Perception**
1.	Schools	Educational Support	No	Empowerment of the community
2.	Clinics	Medical Support	Yes	Improved livelihoods and productivity
3.	Infrastructure development in tourism region	Roads and Infrastructure	Yes	Additional employment

Another perceived attraction in the forest reserves was the natural tranquil environment populated by natural grasses, including thatching grass, natural trees, wild fruits, abundant sand, unused boreholes and wells. This would not only be an attraction in its own right for tourists looking for tranquillity in a natural setting, but would also be an ideal location for the construction of environmentally-friendly accommodations, such as campsites, lodges and mobile safaris. Watering holes and boreholes would also attract wildlife.

4.2. Perceptions of Impacts of Development of Ecotourism: Costs and Benefits

The Botswana Forest Reserves are located within an area that has shortages of arable land. As was discussed earlier under the Study Area, the local community has only around 31% available for communal use. This has resulted in a dire need for more land, since most land (69%) is reserved for conservation [56]. There was a strong feeling, especially among the local communities, that parts of the forest reserves should be designated for communal resettlement to avail more land for construction, agriculture and cattle grazing.

The study revealed that community members had limited access to the forest resources, with a permit, which could only be purchased in the capital city of Gaborone, required for entry. Further concerns arose from the permits being available to people from other parts of Botswana, as this had resulted in people from outside the area camping in the forest reserves for extended periods. They blamed the outsiders for alcohol abuse and crowding during the grass harvesting periods. More importantly, they alluded to the fact that outsiders benefited from their resources. A further issue mentioned by the respondents of the study was the leakage of money being generated by tour operators.

For the development of tourism to take place, facilities, such as roads, electricity, water and shopping facilities, must be in place. The respondents noted that ecotourism development would bring about positive infrastructure developments, such as schools, clinics, roads, electricity and water, to their area.

4.3. Perceptions on Roles of the Private Sector vis-à-vis Local Communities

The perceptions of the respondents of the study were that both the private sector and the local communities should play a role in ecotourism in the Botswana Forest Reserves. In the Sibuyu Forest Reserve heritage sites, the suggestions were that there should be private-public partnerships with the involvement of the community in the interpretation and preservation activities. The Chobe Forest Reserve had three potential sites (Ghoba Pan, Nonotshaa Pool and Kashiba Pool) that could be developed into tented campsites, wildlife viewing and photographic sites, with KALEPA and CECT managing accommodation and ecotourism activities on these sites.

Respondents also perceived cultural tourism to be the preserve of the community in terms of ownership and management, whilst the private sector's role would be in the packaging and marketing of the product. Further perceptions by community members were that the opening up of ecotourism companies would offer employment to local communities within the accommodation and tour guide sectors. Communities acknowledged the importance of the private sector in running successful tourism establishments in Botswana. At different forums, the participants saw the private sector having a role to play in empowering communities to be successful participants in ecotourism businesses. The respondents suggested a partnership between the community and the private sector, where the private sector would go into a joint venture with the community, so as to train and develop the community in business management skills. Other benefits mentioned were the linkages between local farmers and the tourism industry, where local communities would provide food to the ecotourism operators from their farm produce.

5. DISCUSSIONS

Based on the perceptions of study respondents, the opening up of the Botswana Forest Reserves for ecotourism would create opportunities for the poor (see Table 3). Using Mitchell and Ashley's framework [9], the benefits were classified according to three categories; direct effects, secondary effects and dynamic effects.

5.1. Direct Effects

One of the main drivers of pro-poor tourism is the private sector [7]. This was confirmed by the perceptions of the respondents of the study, who supported the existence of a private sector to run tour operation businesses, as well as accommodation facilities in the form of lodges and campsites. This would allow for low-skilled workers to be employed by lodges and campsites as drivers, cleaners and general hands, with the local economy benefitting through the multiplier effect generated by the circulation of their earnings through the community [40]. Whilst the private sector is a huge driver of pro-poor tourism, studies undertaken in the Okavango Delta in Botswana have shown that the private sector imports labour from outside Botswana, and their lodges import a large percentage of building materials from outside Botswana, resulting in leakages from the local economy [57].

There was also a recognition that public-private partnerships could contribute towards the generation of physical and human capital. However, legislation would need to be revised to ensure that capacity building and the development of human capital actually occur, as previous such partnerships have not yielded much success in building human capital. Instead, communities have acted as silent partners who only receive rent from land leased to the private sector [58].

The stakeholders expressed the view that some areas of the forest reserves can have direct benefits to communities. They pointed to the fact that the Chobe Enclave Community Trust (CECT) and the Kasane, Lesoma and Pandamatenga (KALEPA) Trust have already shown an interest in building lodges and camps in some of the forest reserves. Chobe Forest Reserve has three potential sites (Ghoba Pan, Nonotshaa Pool and Kashiba Pool) for ecotourism development focusing on game drives and photographic safaris. In addition, communities have the capacity to develop campsites in the Kasane Forest Reserve, which could be run by the Ngoma, Mabele, Kavimba and Kachikau communities. These developments would increase human capital, with communities taking control of business operations and, in the process, increasing their skill sets.

Studies undertaken in other parts of Botswana have shown that some community-based natural resource management (CBNRM) and SME projects have not resulted in the improvement of the livelihoods of communities. The reasons for failure have been attributed to the insufficient capacity to run such projects, since communities do not have the capital, management and business skills to run a successful safari operation [57,58,59].

5.2. Secondary Effects

The results show two mechanisms of private sector companies sourcing food from local farmers and the use of cultural tourism to alleviate poverty among the local communities. The Chobe District has a total area of 25,000 ha, comprising 245 traditional farms and 21 large-scale commercial farms, which all have the potential to produce a steady supply of food to the tourism establishments in the FRs. The end product would be the sustainable livelihood of communities adjacent to FRs [60].

The informants suggested the development of partnerships and linkages between local communities and commercial tour companies to develop cultural tourism in the villages adjacent to the forest reserves. Some of the suggestions included guided tours of the villages, where tourists would sample local cuisine and partner with local dance groups as part of the tour package. Communities could also be availed opportunities to sell cultural artefacts/souvenirs to tourists. For cultural tourism to succeed, the private sector would need to be more deeply engaged, providing access to their marketing resources, as well as their established international networks. Van Der Duim and Caalders [43] emphasise that, for these linkages to qualify as pro-poor, the private sector must be able to impart basic business skills to the community [61].

5.3. Dynamic Effects

Communities were positive that the development of ecotourism in the Botswana Forest Reserves would result in the improved livelihoods of the communities. Some of the externalities would be the construction of roads to facilitate access to the tourism facilities and hospitals that would support the region. Population growth would then result in increased pressure for the government to build schools, health centres and other supporting facilities.

6. IMPLICATIONS FOR NGOS AND GOVERNMENT

Communities do not have the resources to run successful business operations and would therefore require the support of NGOs, who have in the past also played a pivotal role of providing funding and technical support to pro-poor initiatives. In Namibia, for example, NGOs have provided support for community organisations, like the Namibia Community-Based Tourism Association (NACOBTA), which is a capacity building organisation for community-based tourism organisations [62]. Other examples in South Africa are where the key promoters of the Mehloding Hiking and Adventure Trail project were local NGOs. The role of the NGOs was to identify the project as a possible means of alleviating poverty. They were also responsible for sourcing funding for the project and advertising and marketing the project [37]. The sustainability of the project will, however, depend on the extent to which communities are capacitated to drive the process by deciding what activities they are mostly comfortable with and the market they can confidently serve [61,62].

The government would be required to complement the activities of NGOs and the private sector by developing the legislative framework to guarantee the participation of the poor. It would also be vital that the implementation of pro-poor strategies is monitored to ensure fair play, so that communities were not unnecessarily disadvantaged.

7. CONCLUSIONS

The aim of the study was to investigate the potential for pro-poor tourism development in Botswana's forest reserves as a means for poverty alleviation. The study has shown that Botswana's forest reserves hold great potential for the development of pro-poor tourism and other pro-poor activities. The study agrees with the suggestions of Mitchell and Ashley [9] that the Botswana Forest Reserves can contribute to the alleviation of poverty through ecotourism via three pathways of direct, secondary and dynamic effects. Direct effects that have been discussed in the study include employment creation and the provision of accommodations to tourists. Accommodation would fall under private sector owned and operated, community owned or a partnership between the private sector and communities. The third direct effect would be through the operation of community-owned SMEs specialising in selling crafts and curios/souvenirs. Secondary benefits would arise from the multiplier effect of tourism, whereby villagers would supply agricultural produce to tourism businesses operating in the forest reserves and partnerships in cultural tourism. Lastly, the dynamic effects would be an improved standard of living and sustainable livelihoods.

In conclusion, experiences from elsewhere show that local benefits from protected areas need to exceed the local costs of maintaining the forest in order to reduce the threat of forestry encroachment [11,63,64].

ACKNOWLEDGMENTS

The authors would like to thank Jaap Arntzen and Peter Ruthenberg for the discussions on forest encroachment and the contributions and comments in the initial formulation of this paper, as well as the reviewers and editor for their constructive comments. The study was original funded by the Botswana Tourism Organisation to develop guidelines on the use of forest reserves.

AUTHOR CONTRIBUTIONS

Both authors designed the project. Haretsebe Manwa performed the research and analysed the data; Both researchers wrote the paper on a round robin. All authors read and approved the final manuscript.

CONFLICTS OF INTEREST

The authors declare no conflict of interest.

REFERENCES

1. Government of Botswana. Pverty Eradication Guidelines, 2012. Available online: http://www.gov.bw/Global/OP%20Ministry/Poverty%20Eradication%20Guidelines.pdf (accessed 1 August 2014).
2. Central Statistics Office. *Preliminary results of the 2011 Population and Housing Survey*; Central Statistics Office: Gaborone, Botswana, 2012. [Google Scholar]
3. Sakata, H.; Prideaux, B. An alternative approach to community-based ecotourism: A bottom-up locally initiated non-monetised project in Papua New Guinea. *J. Sustain. Tour.* **2012**. [Google Scholar] [CrossRef]
4. Graci, S. Collaboration and partnership development for sustainable tourism. *Tour. Geogr.***2013**, *15*, 25–42. [Google Scholar] [CrossRef]
5. Matarrita-Cascante, D.; Brennan, M.A.; Luloff, A.E. Community agency and sustainable tourism development: The case of La Fortuna, Costa Rica. *J. Sustain. Tour.* **2010**, *18*, 735–756. [Google Scholar]
6. Ashley, C.; Roe, D. Making tourism work for the poor: Strategies and challenges in South Africa. *Dev. S. Afr.* **2002**, *19*, 61–82. [Google Scholar] [CrossRef]
7. Scheyvens, R. Pro-poor tourism: Is there value beyond the rhetoric? *Tour. Recreat. Res.* **2009**,*34*, 191–196. [Google Scholar]
8. Schilcher, D. Growth *versus* equity: The continuum of pro-poor tourism and neoliberal governance. *Curr. Issues Tour.* **2007**, *10*, 166–193. [Google Scholar]
9. Mitchell, J.; Ashley, C. *Tourism and Poverty Reduction: Pathways and Prosperity*; Earthscan: London, UK, 2010. [Google Scholar]
10. Mbaiwa, J.E.; Sakuze, L.K. Cultural tourism and livelihood diversification: The case of Gcwihaba caves and Xaixai village in the Okavango Delta, Botswana. *J. Tour. Cult. Chang.***2009**, *7*, 61–75. [Google Scholar] [CrossRef]
11. Mbaiwa, J.E.; Stronza, A.L. The effects of tourism development on rural livelihoods in the Okavango Delta, Botswana. *J. Sustain. Tour.* **2010**, *18*, 635–656. [Google Scholar] [CrossRef]

12. Mbaiwa, J.E. Changes on traditional livelihood activities and lifestyles caused by tourism development in the Okavango Delta, Botswana. *Tour. Manag.* **2011**, *32*, 1050–1060. [Google Scholar] [CrossRef]

13. Keitumetse, S.O. The Eco-tourism of Cultural Heritage Management (ECT-CHM): Linking heritage and "environment" in the Okavango Delta regions of Botswana. *Intern. J. Herit. Stud.* **2009**, *15*, 223–244. [Google Scholar] [CrossRef]

14. Lepetu, J.; Alavalapati, J.; Nair, P.K. Forest dependency and its implication for protected areas management: A case study from Kasane Forest Reserve, Botswana. *Intern. J. Env. Res.* **2009**, *3*, 525–553. [Google Scholar]

15. Erskine, L.M.; Meyer, D. Influenced and influential: The role of tour operators and development organisations in tourism and poverty reduction in Ecuador. *J. Sustain. Tour.* **2012**, *20*, 339–357. [Google Scholar] [CrossRef]

16. Spenceley, A.; Meyer, D. Tourism and poverty reduction: Theory and practice in less economically developed countries. *J. Sustain. Tour.* **2012**, *20*, 297–317. [Google Scholar] [CrossRef]

17. Saayman, M.; Rossouw, R.; Krugell, W. The impact of tourism on poverty in South Africa. *Dev. S. Afr.* **2012**, *29*, 462–487. [Google Scholar] [CrossRef]

18. Holden, A.; Sonne, J.; Novelli, M. Tourism and poverty reduction: An interpretation by the poor of Elmina, Ghana. *Tour. Plan. Dev.* **2011**, *8*, 317–334. [Google Scholar] [CrossRef]

19. Sharpley, R.; Naidoo, P. Tourism and Poverty Reduction: The case of Mauritius. *Tour. Hosp. Plan. Dev.* **2010**, *7*, 145–162. [Google Scholar] [CrossRef]

20. Pillay, M.; Rogerson, C.M. Agriculture-tourism linkages and pro-poor impacts: The accommodation sector of urban coastal KwaZulu-Natal, South Africa. *Appl. Geogr.* **2013**, *36*, 49–58. [Google Scholar] [CrossRef]

21. Saville, N.M. Practical strategies for pro-poor tourism: Case study of pro-poor tourism and SVN in Humla West Nepal, Pro-Poor Tourism 2001, Working paper 3. Available online: http://www.propoortourism.inf/documents/Saville2003 WkP3.pdf (accessed on 24 March 2014).

22. Tribe, J. *The Economics of Recreation, Leisure and Tourism*; Elsevier: Oxford, UK, 2005. [Google Scholar]

23. Saarinen, J. Traditions of sustainability in tourism studies. *Ann. Tour. Res.* **2006**, *33*, 1121–1140. [Google Scholar]

24. Hunter, C. Sustainable tourism as an adaptive paradigm. *Ann. Tour. Res.* **1997**, *24*, 850–867. [Google Scholar] [CrossRef]

25. Fox, W. Ecophilosophy and science. *Environmentalist* **1994**, *14*, 207–213. [Google Scholar] [CrossRef]

26. Winters, P.; Corral, L.; Mora, A.M. Assessing the role of tourism in poverty alleviation: A research agenda. *Dev. Pol. Rev.* **2013**, *31*, 177–202. [Google Scholar] [CrossRef]

27. Gronau, W.; Kaufmann, R. Tourism as a stimulus for sustainable development in rural areas: A Cypriot perspective. *Tourismos* **2009**, *4*, 83–95. [Google Scholar]

28. Scheyvens, R. Exploring the tourism-poverty nexus. *Curr. Issues Tour.* **2007**, *10*, 231–254. [Google Scholar] [CrossRef]

29. Farrell, B.; Twining-Ward, L. Seven steps towards sustainability: Tourism in the context of new knowledge. *J. Sustain. Tour.* **2005**, *13*, 109–122. [Google Scholar] [CrossRef]

30. Zapata, M.J.; Hall, C.M.; Lindo, P.; Vanderschaeghe, M. Can community-based tourism contribute to development and poverty alleviation? Lessons from Nicaragua. *Curr. Issues Tour.* **2011**, *14*, 725–749. [Google Scholar] [CrossRef]

31. Sharpley, R. Tourism and development challenges in the least developed countries: The case of The Gambia. *Curr. Issues Tour.* **2009**, *12*, 337–358. [Google Scholar] [CrossRef]

32. United Nations World Tourism Organisation. Available online: http://step.unwto.org/en/content/background-and-objectives (accessed on 20 May 2014).

33. Department for International Development (DFID). *Tourism and Poverty Elimination: Untapped Potential*; DFID: London, UK, 1990. [Google Scholar]

34. Beeton, S. *Community Development through Tourism*; Landlink Press: Collingwood, Australia, 2006. [Google Scholar]

35. Saarinen, J.; Rogerson, C.; Manwa, H. *Tourism, Development and Millennium Development Goals*; Routledge: London, UK, 2013. [Google Scholar]

36. Meyer, D. Pro-poor tourism: From leakages to linkages. A conceptual framework for creating linkages between the accommodation sector and "poor" neighbouring

communities. *Curr. Issues Tour.* **2007**, *10*, 558–583. [Google Scholar]

37. Hill, T.; Nel, E.; Trotter, D. Small-scale, nature-based tourism as a pro-poor development intervention: Two examples in Kwazulu-Natal, South Africa. *Sing. J. Trop. Geogr.* **2006**, *27*, 163–175. [Google Scholar] [CrossRef]

38. Scheyvens, R.; Russell, M. Tourism, land tenure and poverty alleviation in Fiji. *Tour. Geogr.* **2012**, *14*, 1–25. [Google Scholar] [CrossRef]

39. Scheyvens, R. *Tourism for Development: Empowering Communities*; Prentice Hall: Harlow, UK, 2002. [Google Scholar]

40. Ashley, C.; Hayson, G. From philanthropy to a different way of doing business: Strategies and challenges in integrating pro-poor approaches into tourism business. *Dev. S. Afr.* **2006**, *23*, 265–280. [Google Scholar] [CrossRef]

41. Kibicho, W. Community-based tourism: A factor-cluster segmentation approach. *J. Sustain. Tour.* **2008**, *16*, 211–231. [Google Scholar] [CrossRef]

42. Dwyer, L.; Thomas, F. Tourism yield measures for Cambodia. *Curr. Issues Tour.* **2012**, *15*, 303–328. [Google Scholar] [CrossRef]

43. Van der Duim, V.R.; Caalders, J. Tourism chains and pro-poor tourism development: An actor-network analysis of a pilot project in Costa Rica. *Curr. Issues Tour.* **2008**, *11*, 109–125. [Google Scholar]

44. World Travel and Tourism Council. *Botswana; the Impact of Travel and Tourism on Jobs and the Economy*; World Travel and Tourism Council: London, UK, 2007. [Google Scholar]

45. Spenceley, A.; Goodwin, H. Nature-based tourism and poverty alleviation: Impacts of private sector and parastatal enterprises in and around Kruger National Park, South Africa. *Curr. Issues Tour.* **2007**, *10*, 255–277. [Google Scholar] [CrossRef]

46. Torres, R.; Momsen, J.M. Challenges and potential for linking tourism and agriculture to achieve pro-poor tourism objectives. *Progr. Dev. Stud.* **2004**, *4*, 294–318. [Google Scholar]

47. Kirsten, M.; Rogerson, C.M. Tourism, business linkages and small enterprise development in South Africa. *Dev. S. Afr.* **2002**, *19*, 29–59. [Google Scholar] [CrossRef]

48. Trejos, B.; Chiang, L.N. Local economic linkages to community-based tourism in rural Costa Rica. *Sing. J. Trop. Geogr.* **2009**, *30*, 373–387. [Google Scholar] [CrossRef]

49. Lapeyre, R. The Grootberg lodge partnership in Namibia: Towards poverty alleviation and empowerment for long-term sustainability? *Curr. Issues Tour.* **2011**, *14*, 221–234. [Google Scholar] [CrossRef]

50. Burns, P.M. Tourism planning: A third way? *Ann. Tour. Res.* **2004**, *31*, 24–43. [Google Scholar] [CrossRef]

51. Harrison, D. Pro-poor Tourism: A critique. *Third World. Quart.* **2008**, *29*, 851–868. [Google Scholar] [CrossRef]

52. Central Statistics Office. *Wildlife Statistics 2008*; Government Printer: Gaborone, Botswna, 2008.

53. Magole, L.I.; Gojamang, O. The dynamics of tourist visitation to national parks and game reserves in Botswana. *Botswana Note Record* **2005**, *37*, 80–96. [Google Scholar]

54. Norwegian Forestry Society. *Chobe Forests' Inventory and Management Plan*; Ministry of Agriculture: Gaborone, Botswana, 1993. [Google Scholar]

55. Oppermann, M. Triangulation—A methodological discussion. *Int. J. Tour. Res.* **2000**, *2*, 141–145. [Google Scholar] [CrossRef]

56. Lenao, M. Challenges facing community-based cultural tourism development at Lekhubu Island, Botswana: A comparative analysis. *Curr. Issues Tour.* **2013**, *16*, 1–16. [Google Scholar] [CrossRef]

57. Blaike, P. Is Small Really Beautiful? Community-based natural resource management in Malawi and Botswana. *World Dev.* **2006**, *34*, 1942–1957. [Google Scholar] [CrossRef]

58. Sebele, L.S. Community-based tourism ventures, benefits and challenges: Khama Rhino Sanctuary Trust, Central District, Botswana. *Tour. Manag.* **2010**, *31*, 136–146. [Google Scholar] [CrossRef]

59. Jones, T.B. Chobe Enclave, Botswana lessons from CBNRM Project 1993–2002. In *CBNRM Support Programme, Occasional Paper, No.7. IUCN/SVN Support Programme*; The World Conservation Union: Gaborone, Botswana, 2002. [Google Scholar]

60. Ashley, C.; Roe, D.; Goodwin, D. Pro-poor tourism strategies: Making tourism work for the poor. Pro-poor Tourism Report 2001, No 1. Available online: http://www.propoortourism.inf/documents/AshleyetalPPTstrats.pdf (accessed on 6 May 2014).

61. Scoones, I. Sustainable rural livelihoods: A framework for analysis. Institute of Development Studies Working Paper 72, 1998. Available online: http://graduateinstitute.ch/webav/site/developpment/chered/development/mdev/soutienaux.cours (accessed on 30 June 2014).

62. Novelli, M.; Gebhardt, K. Community based tourism in Namibia: "Reality show" or "window dressing"? *Curr. Issues Tour.* **2007**, *10*, 443–479. [Google Scholar] [CrossRef]

63. Mogaka, H.; Simons, G.; Turpie, J.; Emerton, L.; Karanja, F. *Economic Aspects of Community Involvement in Sustainable Forest Management in Eastern and Southern Africa*; International Union for Conservation of Nature and Natural Resources, Eastern Africa, Regional Office: Nairobi, Kenya, 2001. [Google Scholar]

64. Dudley, N.S.; Mansourian, S.; Stolton, S.; Suksuwan, S. *Safety Net: Protected Areas and Poverty Reduction*; World Wide Fund for Nature and Equilibrium: Gland, Swizerland, 2008.

6

CHAPTER

Tourism and Sustainability in the Evaluation of World Heritage Sites, 1980–2010

Vaughn Schmutz [1,*] **and Michael A. Elliott** [2]

[1] *Department of Sociology, University of North Carolina at Charlotte, Charlotte, NC 28269, USA*

[2] *Department of Sociology, Anthropology & Criminal Justice, Towson University, Towson, MD 21252, USA*

ABSTRACT

At present, there are myriad concerns about tourism and sustainability at cultural and natural world heritage sites. Based on an analysis of 811 evaluations written between 1980 and 2010 by two official advisory bodies to the World Heritage Committee, this paper charts the timing and extent to which such concerns have become central to assessing the value of heritage sites. We find that, over time, issues related to tourism and sustainability expanded considerably in quantity and variety, and recommendations for managing and developing sustainable tourism became a routine feature of site evaluations. Despite the growing prevalence of such concerns, the conceptualization of sustainable tourism and related recommendations provided by the advisory experts remain somewhat ambiguous. Furthermore, our findings reveal regional disparities in the degree to which tourism is seen as a threat to the sustainability of heritage sites and in the likelihood that a state is considered a model of sustainable tourism.

Keywords: world heritage; sustainable tourism; world society; cultural wealth

1. INTRODUCTION

In 1972, the Convention concerning the Protection of the World Cultural and Natural Heritage, or World Heritage Convention, was adopted at the seventeenth session of UNESCO's General Conference [1]. Building on key institutions in world society [2], the Convention's aims to identify, protect and preserve "cultural and natural heritage of outstanding universal value" have rapidly diffused throughout the world and achieved near universal acceptance in the decades since. Indeed, when the Bahamas ratified in 2014 it became the 191st State Party to adhere to the World Heritage Convention, furthering its status as the most widely accepted UNESCO convention [3]. Yet just as the World Heritage List has expanded to include over one thousand cultural and natural heritage sites, the concerns and challenges associated with successfully preserving world heritage have multiplied as well [4]. Among those concerns are the sustainability of an ever-expanding number of inscribed sites [5] and the social, economic, and environmental impacts of tourism on cultural and natural properties [6,7,8]. Although the World Heritage Convention predates the concept of sustainable tourism, Labadi [9] suggests that it is implied in Article 4, which places responsibility with each state party to ensure the "presentation and transmission to future generations" [1]. This brief statement highlights the difficult balance between preserving world heritage and making it accessible to the public.

Today, such concerns have been formally incorporated into UNESCO's World Heritage and Sustainable Tourism Program [10], as well as sustainable development initiatives [11], which seek to balance the interests of world heritage's diverse stakeholders. While efforts to promote sustainable tourism have inspired a great deal of scholarly analysis, debate, and even skepticism [12,13,14,15], this paper focuses on how and when issues of sustainability and tourism emerged in the world heritage arena as well as how concerns about these issues have changed over time. In particular, we highlight the role of two official advisory bodies—the International Council on Monuments and Sites (ICOMOS) and the International Union for the Conservation of Nature and Natural Resources (IUCN)—in assessing and articulating concerns about sustainability and tourism through an analysis of their evaluations of over 800 cultural and natural sites nominated for inscription on the World Heritage List. Our approach is inspired, in part, by institutional [16] and "cultural wealth" approaches [17] to global analysis in sociology, which we briefly describe below.

Many observers have understandably criticized heritage tourism as either a profit-making tool of the tourism (or heritage) industry; a means of identity construction and self-aggrandizement for nation-states that reflects elite interests [18,19,20]; or a Eurocentric imposition by Western countries and international agencies [21,22,23]. Among other things, such efforts to commoditize, politicize, or universalize heritage are seen as a threat to the authenticity of cultural and natural properties [24,25,26]. Yet scholars have also shown that inscription on the World Heritage List is neither a guaranteed boon to tourism [27,28,29] nor a process that is always driven by elites [30]. Other studies highlight the problems with conceptualizing the tourism industry as a monolithic entity [31] and demonstrate the role of local influences in heritage outcomes and interpretations [32,33]. Thus, while power imbalances are clearly evident in the world heritage arena, a wide range of stakeholders potentially shape the development, interpretation, and inscription of cultural and natural sites [34]. Although the cultural and natural heritage experts who assess the value of nominated world heritage sites are among these stakeholders, they are rarely the subjects of scholarly analysis [5] despite theoretical reasons for doing so.

World society theory, an institutional approach to global analysis in sociology [16], emphasizes the influence that a variety of actors exert at different levels (e.g., local, national, supranational, and global) in generating and reproducing global structures and processes. The notion of a world polity is invoked to describe a highly diffuse authority structure, without geographical boundaries or an administrative center, which has become increasingly authoritative relative to nation-states in propagating rules and setting standards of behavior in many spheres of social life [35]. While a wide range of actors (e.g., intergovernmental organizations, nation-states, universities, corporations, voluntary associations, individuals) can exercise legal sovereignty in the world polity, non-governmental organizations (NGOs) have become especially influential since the 19th century [36,37], shaping a wide range of transnational movements including those promoting world heritage [2], environmentalism [38,39,40], and sustainable tourism [41].

The recently articulated "cultural wealth" perspective in global sociology, which draws heavily on the work of Pierre Bourdieu, focuses on the role of reputational attributes and cultural products in establishing the global status and prestige of a nation [17]. Just as individuals use their "cultural capital" (*i.e.*, cultural knowledge, tastes, styles of interaction), in Pierre Bourdieu's [42] formulation, to establish and reproduce their social position, nations draw on their cultural wealth to maintain their status in the global arena. Furthermore, cultural wealth can be converted into economic and political advantage by making a country more attractive to foreign investors or tourists, a process that reproduces global inequalities between states. While international prizes received by its citizens or artworks exhibited in prestigious museums are indicators of cultural wealth, many scholars taking this approach have focused on the quantity and quality of a nation's natural and cultural heritage sites as a source of "symbolic value" [43,44,45,46,47]. From this perspective, states engage in "international impression management" [48] to convey their status to a world audience. Yet Bourdieu's theory also points to the key role of cultural intermediaries in legitimating the reputations of high status actors in fields of cultural production [49]. In Bourdieu's terms, INGOs like ICOMOS and IUCN act as "consecrating agents" who verify the cultural and natural heritage sites that exemplify outstanding universal value.

Therefore, we focus our analysis on the evaluations produced by ICOMOS and IUCN due to their influence in setting standards for tourism and sustainability at inscribed sites as well as their role in certifying their universal value. From a world society perspective, the evaluations can provide insights into the extent and timing of the emergence, growth, and institutionalization of concerns about tourism and sustainability in the assessment of world heritage sites as well as how such concerns have changed over time. From a cultural wealth perspective, the evaluations are relevant to understanding how well different states are able to convey the value of their heritage and their competency in managing tourism in a sustainable manner through their nomination dossiers. A state's ability to give the right "impression" to ICOMOS or IUCN is not only important for getting a proposed heritage site inscribed but also indicates how well they will be able to convert their cultural wealth into economic and other forms of advantage. Taken together, our analysis highlights the institutionalization of tourism and sustainability concerns as well as disparities in the perception of touristic threats in the evaluations of heritage sites. Our findings also have practical implications with respect to the role of the advisory bodies in providing specific recommendations that promote the sustainable development and management of heritage tourism.

2. MATERIALS AND METHODS

2.1. Materials: Advisory Body Evaluations of Nominated Sites

Evaluations of nominated sites are a central component in the inscription process (see Figure 1). In order to nominate a property for inclusion on the World Heritage List, a country must first sign the World Heritage Convention. Next, the State Party compiles an inventory of its most prized natural and cultural heritage sites, called a Tentative List [50], which represents possible submissions for inscription in the next five to ten years. After creating a Tentative List, a State Party may submit a nomination file for a particular site. Today, nomination files are extensive; they typically include a geographical identification of the site with coordinates and detailed maps, information about its history and development, statements of integrity and authenticity, a comparative analysis with similar properties that have achieved inscription, specific details about conservation and management efforts (e.g., legislative and other regulatory measures, concrete indicators to measure success), and so on. The required contents of the nomination dossiers, which have significantly expanded in complexity over time [2,4,9], are outlined in the *Operational Guidelines for the Implementation of the World Heritage Convention* [51]. Then, each nominated property receives an independent inspection by either ICOMOS, in the case of cultural sites, or IUCN, in the case of natural sites. The results of this inspection are documented in an official "advisory body evaluation" that includes particular details about the site and expert comments on its eligibility for inclusion on the World Heritage List. The evaluations contain recommendations to the State Party and to the World Heritage Committee, which makes the final decisions with regard to inscription.

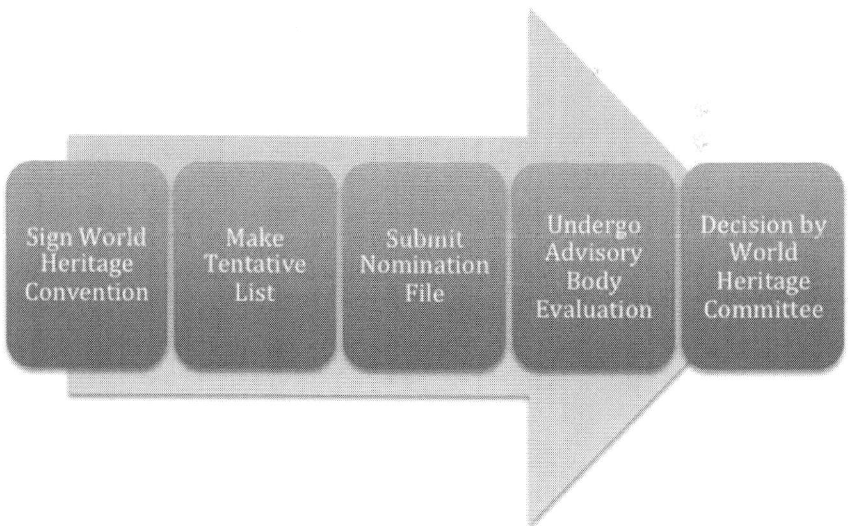

Figure 1. World heritage site nomination process.

Both ICOMOS and IUCN have played a prominent role in the development of the world heritage movement [2] (ICOMOS emerged out of the Second

Congress of Architects and Specialists of Historic Buildings in 1964 and operates in a variety of capacities to develop and disseminate knowledge about preservation standards and techniques. Today, ICOMOS comprises a global network of experts from a variety of specializations (e.g., architecture, history, archaeology, geography, anthropology, engineering, town planning, *etc.*), who work to promote "the application of theory, methodology, and scientific techniques to the conservation of architectural and archaeological heritage" [52]. IUCN was founded in 1948 to promote international cooperation in natural conservation efforts. Early on, its role included drafting the first UN List of Parks and Equivalent Reserves in 1959 as well as a 1968 proposal to combine the conservation of cultural and natural sites [53].). As permanent advisory bodies, ICOMOS and IUCN play a central role in the process of evaluating sites and, despite some recent controversies, the World Heritage Committee has historically followed their recommendations with regard to inscription fairly closely [54,55]. Our data includes every evaluation written by ICOMOS and IUCN for inscribed cultural and natural sites from 1980 to 2010 (N = 811), all of which were retrieved from UNESCO's World Heritage Centre website [56]. In order to enhance our comparison of sites that are either distinctly cultural or distinctly natural, we exclude the few mixed sites (N = 25) that were inscribed during this time period. Mixed sites contain both cultural and natural features and are typically evaluated by both ICOMOS and IUCN. Table 1 provides a year-by-year breakdown of the evaluations in our sample by site type.

Table 1. Cultural and natural site evaluations analyzed, by year and type.

Year	Cultural (ICOMOS)	Natural (IUCN)	Year Total
1980–1985	113	37	150
1986–1990	74	21	95
1991–1995	105	27	132
1996–2000	180	36	216
2001–2005	99	21	120
2006–2010	77	21	98
Site total	648	163	811

2.2. Methods: Analysis of Advisory Body Evaluations

To prepare the evaluations for analysis after they were downloaded as pdf files, they were processed using optical character recognition (OCR) software and converted back to pdf files. Finally, the evaluations were imported into MAXQDA, a text-based data analysis software that facilitates both quantitative and qualitative content analysis. First, we employed the MAXDictio add-on to the software program to construct a dictionary of key terms and to conduct word counts of the evaluations. In addition to producing total word counts of each evaluation, we used this function to identify all evaluations that contained some direct mention of tourism (including tourism, tourist, and other variants) or of sustainability (including sustainable, unsustainable, sustainably, and other variants.). A strength of this approach is that it enables us to broadly identify when discussions of tourism and sustainability emerged in cultural and natural site evaluations as well as the general extent to which it has become a routine feature of the evaluations over time. However, it is limited in the sense that it does not tell us exactly what was discussed, regarding tourism and sustainability, and how these discussions change over time.

Therefore, we next conducted a detailed qualitative content analysis of a subset of the evaluations that discussed tourism. To give us a sense of how such concerns have changed over time, we focused on all the evaluations from the 1980s (n = 44) and all evaluations from 2006 to 2010 (n = 90) that discussed tourism. Our coding focused on determining for each evaluation: whether or not ICOMOS or IUCN identified tourism as a threat to the site; whether ICOMOS or IUCN expressed positive sentiments about tourism (e.g., expressed optimism about the potential for tourist development, praised the state party for sustainable management of tourism, *etc.*); and whether ICOMOS or IUCN made specific recommendations with regard to sustainable tourism. Finally, we used this information to identify trends over time as well as to make comparisons across site type (cultural *vs.* natural) and region.

3. RESULTS

Below, we first present results of the word count analysis to demonstrate the timing and extent of discussions of tourism and sustainability in the advisory body evaluations produced by ICOMOS and IUCN. Then we provide additional details based on our qualitative coding of the evaluations that contain mentions of tourism.

3.1. Word Count Results

First, we provide some contextual information about changes in the advisory body evaluations, based on previous research [4]. Overall, the evaluations of cultural and natural sites grow considerably longer over time. Initially, evaluations are only a few hundred words in length but steadily expand over time to several thousand words. Although natural site evaluations are somewhat longer through the 1980s and early 1990s, cultural site evaluations catch and then surpass natural site evaluations in length in recent years (*i.e.*, over 8000 words for cultural sites, on average; over 6000 words for natural sites, on average). This general expansion of the evaluations indicates that they have come to include a much wider range of information and topics of concern over time.

Figure 2 presents the percent of cultural and natural site evaluations that contain direct discussion of tourism over time. Initially, there is a noticeable gap between natural and cultural site evaluations as ICOMOS rarely raised tourism concerns, while IUCN discussed tourism in most evaluations by 1983 and continued to do so thereafter. Particularly during the 1990s, however, cultural and natural site evaluations converged. By 2006, over 80% of cultural site evaluations and 100% of natural evaluations discuss tourism and nearly all evaluations do the same thereafter, which indicates that tourism-related issues have become firmly institutionalized in the assessment of world heritage. By contrast, discussion of sustainability is much slower to emerge, particularly in cultural site evaluations (see Figure 3). After 1999, it begins to appear more often and gradually approaches the proportion of natural site evaluations that directly address sustainability. At the end of the time period we study, about half of all evaluations produced by both ICOMOS and IUCN contain some discussion of sustainability.

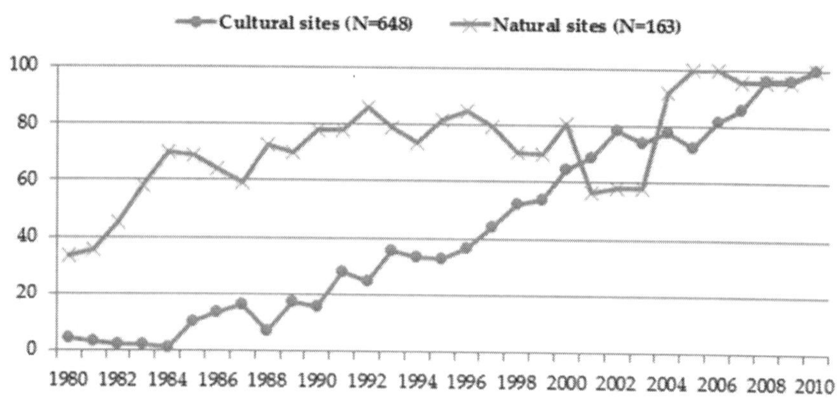

Figure 2. Percent of cultural and natural site evaluations that discuss tourism.

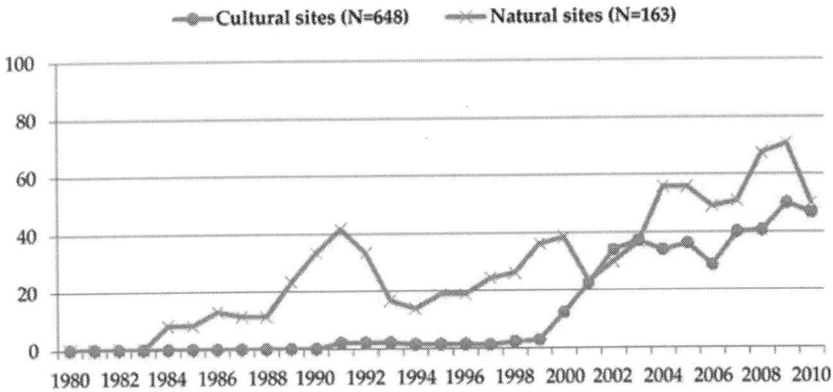

Figure 3. Percent of cultural and natural site evaluations that discuss sustainability.

Based on the word count patterns, it is clear that sustainability is not typically associated with discussions of tourism until recent years, particularly in the evaluations of cultural sites. Across all evaluation types, the vast majority (79.4%) of instances in which the two concepts are linked occur after 2000, despite the fact that sustainable tourism had already been the subject of considerable international organizing since the late 1980s and early 1990s [41]. In total, 102 of the 811 evaluations (12.6%) discuss both sustainability and tourism together. Although the term "sustainable tourism" itself is relatively rare, mentions of sustainable development or sustainable management are sometimes linked to tourism-related concerns. Nearly one in four natural site evaluations (23.9%) contain terms related to both sustainability and tourism, but only about one in ten cultural site evaluations do the same (9.7%). If we focus on the latest five years of our study (2006–2010), 11 out of 20 (55.0%) natural and 29 out of 70 (41.4%) cultural evaluations include both tourism and sustainability, which indicates that it is becoming fairly common by the end of our study period. To some extent, mentions of tourism and sustainability may reflect changes in the*Operational Guidelines* [51] that outline the issues States Parties must address in their nomination dossiers. Therefore, we turn to our qualitative coding results to focus on changes in the ways that the advisory bodies assess the threats and opportunities associated with tourism and sustainability as well as the relevant recommendations they provide.

3.2. Qualitative Coding Results

The results of the qualitative coding provide additional insights into the patterns described above. Although the discourse of sustainability is rarely invoked in the 1980s evaluations, discussion of tourism gradually increases for cultural sites during this decade and becomes the norm for natural sites. Yet even when tourism is discussed, it is typically fairly brief and often more descriptive than evaluative. For instance, the ICOMOS evaluation of India's Group of Monuments at Hampi simply mentions one monument as being a "favorite of tourists today" and refers to a plan for developing "touristic and scientific infrastructure" [57] (p. 2). Likewise, IUCN writes that the waterfalls in Iguazu National Park are "one of the principal tourist attractions in Argentina" and notes that there are plans for additional tourist facilities, but does not describe this as a concern [58] (p. 2). Even for some sites that attract a large number of visitors, tourism is not described as a threat. For example, "touristic development" is described by ICOMOS as an integral part of urban Rhodes in Greece, "desirable or not" [59] (p. 3); and any concerns about the 680,000 yearly visitors to the Great Smoky Mountains in the USA are downplayed by the IUCN evaluation, which notes that the tourist infrastructure has mostly been developed outside the Park and it provides economic benefit to people in the surrounding area [60]. In some cases, brief advice is provided by the advisory body, whether or not a specific threat is identified, such as when ICOMOS recommended that no "tourist facilities" be constructed within view of the Temple of Apollo Epicurius at Bassae in Greece [61] (p. 3). As in the examples offered here, most evaluations in the 1980s are fairly equivocal and do not explicitly refer to tourism as a threat to heritage.

In total, only 11 of the 44 cultural and natural evaluations (25.0%) characterize tourism as a threat, either currently or potentially, to the site under review. A few instances elicit strong concern from the advisory bodies, such as ICOMOS' recommendation, with regard to the Great Zimbabwe National Monument, to postpone "installation of tourist facilities which are expensive and dangerous" until the archaeological areas are properly excavated [62] (p. 3). Similarly, a review of the Chan Chan Archaeological Zone in Peru leads ICOMOS to question the restoration and development projects underway, which it sees as efforts "to satisfy the demands of mass tourism" [63] (p. 3). Among the examples from natural site evaluations, IUCN describes Tanzania's Kilimanjaro National Park as threatened by high "tourism intensity", which results in litter and challenges of facility maintenance [64] (p. 3). Yet most of the sites for which tourism is seen as a threat include somewhat vague and usually succinct statements. Paphos in Cyprus, for instance, is simply said to be threatened by "commercial (touristic) developments" [65] (p. 2), while tourism is described as a "disturbance" in Nepal's Royal Chitwan National Park, ranking just behind the threats from fire and grass-cutting by locals [66] (p. 2).

By comparison, nearly as many cultural and natural evaluations in the 1980s—10 out of 44 (22.7%)—describe tourism as having a positive relationship with the site, either as an indicator if its outstanding universal value, a boon to the local economy, or due to exemplary management. Ksar of Ait-Ben-Haddou in Morocco, for instance, is said to possess every indication that it belongs on the World Heritage List "because of the celebrity it enjoys among tourists" [67] (p. 2). Likewise, the development of a new "tourist complex" at Tipasa is credited with bringing "new life to the ruins", although ICOMOS recommends that the Algerian government place strict controls on any future tourist facilities [68] (pp. 2–3). The Historic Center of Evora in Portugal is described as one of the major tourist attractions in the region, but ICOMOS emphasizes that the "museum city has retained all of its traditional charm" [69] (p. 1). Among natural sites, IUCN welcomes efforts to promote tourism in Cote d'Ivoire's Comoe National Park and even suggests expanding the Park's boundaries to enhance the "ecological and touristic value of the property" [70] (p. 2). Plans to develop the park infrastructure and partner with the private sector tourism industry in Canada's Gros Morne National Park are encouraged by IUCN because they will "diversify and increase employment opportunities for local communities" [71] (p. 3). In short, when evaluations in the 1980s made mention of tourism it was typically brief, ambiguous in its assessment, and rarely made specific recommendations. Table 2 summarizes the coding results and also reports them by region.

Table 2. Summary of coding results for 1980s site evaluations that discuss tourism.

	Threat	No Threat/Unclear	Positive	Recommendation
TOTAL % (*N* = 44)	25.0% (11)	52.3% (23)	22.7% (10)	22.7% (10)
AFR % (*N* = 12)	16.7% (2)	41.7% (5)	41.7% (5)	25.0% (3)
APA % (*N* = 9)	11.1% (1)	88.9% (8)	0.0% (0)	22.2% (2)
ARB % (*N* = 3)	0.0% (0)	33.3% (1)	66.7% (2)	33.3% (1)
EUR % (*N* = 15)	33.3% (5)	53.3% (8)	20.0% (3)	20.0% (3)
LAC % (*N* = 5)	60.0% (3)	40.0% (2)	0.0% (0)	20.0% (1)

Abbreviations refer to UNESCO Regions (AFR = African States; APA = Asia and the Pacific; ARB = Arab States; EUR = Europe and North America; LAC = Latin America and the Caribbean).

The regional breakdown of the coding results indicates that Latin American/Caribbean (LAC) sites as well as the European/North American sites (EUR) have a higher than average proportion of evaluations that describe tourism as a threat. Yet EUR sites, along with African (AFR) sites, also have a higher than average proportion of ambiguous evaluations in which tourism is not identified as a specific threat. AFR sites have a high proportion of positive comments about tourism, several of which involve praise from IUCN for having well-managed national parks. Arab states (ARB) also receive positive assessments about tourism in 2 out of its 3 inscribed sites during the 1980s, while sites in Asia/Pacific Islands (APA) and LAC regions contain no positive comments about tourism. Across all regions, a similar proportion of sites receive specific recommendations from the advisory bodies ranging between 20.0% for EUR and LAC sites to 33.3% for ARB sites.

Whereas tourism is discussed to a limited extent and often in ambiguous ways in the 1980s, concerns about tourism become a more consistent aspect of advisory body evaluations over time, especially after the issue of sustainable tourism became an institutionalized principle of world heritage discourse. Just as the evaluations expand in length and complexity over time, so do the assessments of tourism-related issues. In our qualitative analysis of cultural and natural site evaluations written from 2006 to 2010 ($N = 90$), we continued to code for direct mentions of tourism as a threat; positive sentiments about tourism or its management at the site; and descriptive or ambiguous comments about tourism that do not provide a negative or positive assessment of tourism. Yet due to the increased amount and complexity of comments about tourism, some combination of these views could appear in the same site evaluation. So, for a few sites, the evaluations may identify tourism as a threat but also include a positive assessment of tourism development plans or management. Also, we coded instances in which ICOMOS or IUCN make specific recommendations with regard to the development, management, interpretation, or other aspects of tourist activity at a site.

Similar to the site evaluations written in the 1980s, most evaluations (63.3%) do not identify tourism as a threat or are at least equivocal about its relationship to the site from 2006 to 2010. During most of this time period, "tourism pressures" is an institutionalized subheading in ICOMOS evaluations under the broader heading "factors affecting the property", but many evaluations explicitly state that tourism is not much of a concern in this section. What is much different during this time period, however, is that ICOMOS and IUCN are much more likely to make specific recommendations—57.8% of the time, compared to 22.7% of the time in the 1980s—even when tourism is not described as a significant threat to the site currently. In such cases, the appropriate advisory body often recommended a formalized management plan or strategy to prepare for the *possibility* of increased tourism, even though it was not apparent at the time. Following world society theory, we think such examples highlight the fact that concerns over tourism are not merely a functionalist response to "actual" increasing threats, but are often the ritual expression of concerns that have widely diffused and become formally institutionalized in official documents and decrees [16].

Another difference, relative to the 1980s, is that formal recommendations, many of which focus on future tourism plans, are most common in the evaluation of sites from less developed countries during 2006 to 2010. As shown in Table 3, evaluations of sites in African (66.7%), Asian (74.2%), and Arab (83.3%) states are much more likely to contain such recommendations compared to sites in European/North American (42.4%) and Latin American (36.4%) states during this period. For example, in their evaluation of the Bisotun archaeological site in Iran, ICOMOS made no mention of tourism as a threat but recommended that the issues of visitor reception and the presentation of the site needed to improve in the future. What is more, they requested than any future plans regarding these issues receive the prior approval of the World Heritage Committee: "ICOMOS requests the State Party to report to the World Heritage Committee on the design and construction of access routes to the site and the development and design of eventual visitor facilities, before such works are undertaken" [72] (p. 74). Given that no specific threats were identified in the evaluation, this seems to be a fairly heavy-handed recommendation.

Table 3. Summary of coding results for 2006–2010 site evaluations that discuss tourism.

	Threat	No Threat/Unclear	Positive	Recommendation
TOTAL % (*N* = 90)	37.8% (34)	63.3% (57)	14.4% (13)	57.8% (52)
AFR % (*N* = 9)	44.4% (4)	55.6% (5)	0.0% (0)	66.7% (6)
APA % (*N* = 31)	54.8% (17)	51.6% (16)	9.7% (3)	74.2% (23)
ARB % (*N* = 6)	33.3% (2)	66.7% (4)	0.0% (0)	83.3% (5)
EUR % (*N* = 33)	24.4% (8)	72.7% (24)	21.2% (7)	42.4% (14)
LAC % (*N* = 11)	27.3% (3)	72.7% (8)	27.2% (3)	36.4% (4)

Abbreviations refer to UNESCO Regions (AFR = African States; APA = Asia and the Pacific; ARB = Arab States; EUR = Europe and North America; LAC = Latin America and the Caribbean).

Likewise, in its 2007 evaluation of the Red Fort Complex in India, ICOMOS expressed concern about the lack of a visitor management strategy to distribute tourists and recommended that an "integrated approach to research, conservation, monitoring, tourism, education and local social values" be included in their Comprehensive Conservation Management Plan [73] (p. 103). In its 2008 evaluation of the Socotra Archipelago in Yemen, IUCN noted that Socotra was an isolated site with few tourists. Nonetheless, they recommended very specific objectives to be included in an Ecotourism Master Plan that:

(a) maintains the current focus on low key, nature based tourism, based on the appreciation of natural values; (b) includes a carrying capacity assessment to guide tourism development ...; (c) provides for direct and adequate financial contributions from tourism to ... conservation and community development ...; (d) closely involves the Yemeni General Tourism Development Authority and Tourism Promotion Board; (e) considers options for engaging in partnerships with environmentally sensitive private sector; and (f) addresses the lack of trained local tourist guides and literature

[74] (p. 9).

And in Iran, once again, the 2010 evaluation of the Tabriz Historic Bazaar Complex by ICOMOS found that there was "no pressure from tourism" yet it was recommended that training programs on sustainable tourism be developed for owners and managers to make sure they were prepared to tackle tourism issues [75] (p. 138).

At the same time, site evaluations more often described tourism as a current or imminent threat from 2006 to 2010 (37.8% of the time) compared to the 1980s (25.0% of the time). Even more striking is the increasing variety of forms such threats took, according to ICOMOS and IUCN. Again, sites in less developed countries were more likely to be seen as threatened by tourism with the highest rates evident in Asian (54.8%) and African (44.4%) sites during this period. Not surprisingly, extensive human contact via tourism could be a threat when it degraded material structures or natural environments. For example, in its 2009 evaluation of The Sulaiman-Too Sacred Mountain in Kyrgyzstan, ICOMOS highlighted the persistence of vandalism and negligence, writing: "Uncontrolled visitor access for many years resulted in a damage to petroglyphs found in the lower parts of the first and second peaks: a number of modern graffiti, mainly painted, but also those engraved and/or pecked, cover many rock surfaces" [76] (p. 25).

Similarly, the construction of tourist accommodations could be viewed as excessive and indiscriminate when it degraded the site itself. For example, in its 2008 evaluation of the Sacred Mijikenda Kaya Forests in Kenya, ICOMOS recommended that one of the forests (Kaya Kinondo) be removed from the nomination altogether due to "severe threats" from tourism development, which had replaced forest land with visitor accommodations and threatened traditional ways of life [77] (p. 4). In this case, as in others, eco-tourism or sustainable tourism efforts were a common solution. In the evaluation of the Monarch Butterfly Biosphere Reserve in Mexico in 2008, IUCN found that tourist accommodations did not degrade the site or directly harm the butterfly colonies but, rather, were unsightly. They write: "Current tourism impacts relate not so much to the butterfly colonies, but rather to the area's natural beauty. Most of the existing tourism infrastructure has been developed by local communities without considering visual or environmental impacts, and this detracts in a major way from the visual integrity of the sites that are visited by tourists" [78] (p. 97).

In some cases, the threat of tourism extended to traditional ways of life that are integrated into the physical environment of the site. For example, in its 2008 evaluation of the Fujian Tulou—a group of large, communal clan houses in China—ICOMOS identified imminent threats from tourism that could degrade the site and the surrounding environment, which could then affect the lifestyle of its inhabitants. "It is recognized ... that too many visitors may adversely impact the structures, especially the wooden floors, which will affect the traditional way of life and culture. The creation of tourist facilities will, furthermore, have a negative impact on the environment" [79] (p. 26). Adding to this, ICOMOS argued that the physical presence of tourists themselves could diminish the authenticity of the site. "The *tulou* exist in a fragile environment that will come under additional pressure with the expected growth in tourism. Part of their value lies in their relationship with agricultural activities carried out in the surrounding landscape. If all the *tulou*became museums there could be a significant loss in authenticity" [79] (p. 26). Thus, while tourist pressures tend to threaten material structures or environmental landscapes, in this case the very act of visitors being present (and observing) threatens the traditional behavior of those who still inhabit these structures, which is part-and-parcel of their outstanding universal value. Overall, the threats of tourism appear more often in cultural and natural site evaluations in later years and they are described in much greater detail and variety compared to earlier evaluations.

Given the increased identification of tourism threats, it is not surprising that fewer site evaluations express positive sentiments with regard to the relationship between tourism and heritage sites between 2006 and 2010. During the 1980s, nearly one in four site evaluations (22.7%) expressed some positive aspect of tourism whereas only about one in seven (14.4%) do the same in the later time period. With the exception of three positive mentions in Asian site evaluations (*i.e.*, 9.7% of the time), all positive expressions appear in sites from the EUR (21.2%) and LAC (27.2%) regions. As a positive attribute, tourism was viewed as something that could add value to a potential world heritage site, particularly if it was already well managed by the nominating State Party. For example, in its 2006 evaluation of the Old Town of Regensburg in

Germany, ICOMOS noted that the site was well managed and "the city is well prepared for tourism" [80] (p. 104). In the same year, ICOMOS' evaluation of Le Strade Nuove in Genoa noted that the city had much improved its capacity to manage tourism and gave no specific recommendations about tourism development [81]. ICOMOS' evaluation of the Jeju Volcanic Island and Lava Tubes in South Korea in 2007 acknowledged a high number of tourists to the site but considered it well managed and not beyond its "carrying capacity" and also applauded the site as an exemplar of the involvement of civil society in world heritage due to the high level of public awareness in Jeju Province about the site, its value, and potential inscription [82] (p. 33). Despite 650,000 visitors per year, ICOMOS concluded in 2010 that the Episcopal City of Albi in France was well managed and that pressures were well understood. Nonetheless, they recommended the need for tourist signs and for "tourism monitoring indicators" to be defined and implemented [83] (p. 243). Finally, ICOMOS found no issues with tourism at the Seventeenth-Century Canal Ring Area of Amsterdam in its 2010 evaluation, despite nearly 5 million hotel nights by tourists in 2007, because facilities were adequate and growing. Tourism was also seen as an important part of the economic growth of this city [84].

4. DISCUSSION

Taken together, the results of our analysis of the advisory body evaluations offer a number of insights, both practical and theoretical. As we argued in the introduction, ICOMOS and IUCN are influential stakeholders in the world heritage arena, although different theoretical perspectives conceptualize their roles in somewhat divergent ways. In our view, foregrounding the part that these advisory organizations play in assessing and articulating the relationship between tourism and the value of world heritage sites complements and contributes to the extant scholarship on this topic. Among other things, it demonstrates that tourism-related issues have joined a panoply of concerns that heritage experts must take into account when appraising the universal value of a cultural or natural property. Yet given the growing variety of sites nominated for inscription and the increasing complexity of the evaluations, it is small wonder that discussion of and recommendations regarding sustainable tourism remain fairly ambiguous in this early stage of institutionalization.

Indeed, some have noted the uneasy relationship between sustainable tourism and development, suggesting that UNESCO appears to simultaneously embrace it with one hand and shun it as a threat to heritage with the other [9,12]. Some of our findings certainly resonate with the uncertainty and ambiguity with which ICOMOS and IUCN address tourism in their site evaluations. In attempting to transform abstract concepts (e.g., outstanding universal value, sustainable tourism) into concrete practice, it is not entirely surprising that straightforward recommendations and uniform interpretations are hard to come by. Others have argued that heritage professionals were highly antagonistic to tourism in decades past but have become more favorable to the tourism industry in recent years [5]. Our analysis suggests that this generalization may not extend to the experts that evaluate sites for World Heritage List inscription, as they appear to have become more concerned about the threats of tourism and less likely to highlight its positive aspects by the end of our study period. Given that site evaluations in less-developed countries are more likely to include formal recommendations regarding tourism plans, evaluators should be particularly sensitive to the balance between development and sustainable tourism in these countries, which may rely heavily on tourist revenues yet lack some of the bureaucratic and technical capacity required to ensure robust management plans.

On a practical level, these findings highlight the ongoing challenges of clarifying what constitutes sustainable tourism and how it can be applied to cultural and natural properties of such remarkable diversity. Gullino, Beccaro and Larcher [85] similarly note a lack of specific models, rules or definitions associated with UNESCO's call for sustainable management plans and they demonstrate the challenges this presents, specifically for rural world heritage sites. The difficulty of providing clear recommendations to States Parties may also contribute to some of the frustrations that have led to pointed criticism of the advisory bodies by delegates to the World Heritage Committee in recent years [55]. Next, we highlight the relevance of our findings to two approaches to global analysis in sociology, world society theory and the cultural wealth perspective.

4.1. World Society Theory

The expansion of discourse about sustainable tourism in the site evaluations follows a pattern that has been frequently documented by world society scholars in other domains of global activity. As ideas about human rights [86], corporate social responsibility [87], or government transparency [88], for example, become widely diffused and taken for granted, they are increasingly applied and enacted in various institutional settings around the world. Similarly, as sustainable tourism becomes a culturally legitimate principle in world society, it becomes a routinized feature of cultural and natural site evaluations. This is especially evident from the fact that specific recommendations—for example, to create a formal plan for the sustainable development and management of tourism—become commonplace even for sites that receive few tourists or where tourism is not perceived as a threat to their integrity. Consistent with world society theory, specialized INGOs like ICOMOS and IUCN exert considerable authority in defining and propagating standards of sustainable tourism in the world heritage arena, based on their scientific and technical expertise. Future research from this perspective could also explore the global emergence and diffusion of tourism- and sustainability-related concerns outside the world heritage arena (e.g., in the United Nations, at world conferences, and among other prominent INGOs) and how this may have affected the timing of their expansion in the recent advisory body evaluations of ICOMOS and IUCN.

4.2. Cultural Wealth of Nations

At the same time, our findings resonate with aspects of the recent cultural wealth perspective in global sociology and are suggestive of ways to develop it further [17]. As described in the introduction, this approach focuses on the ways that nations draw on their stocks of cultural capital to enhance their status and reproduce other forms of advantage (e.g., economic, political). Countries that are endowed with cultural wealth are better able to manage the impressions that others in the world system hold of them and countries can engage in impression management to emulate the cultural styles of high-status nations [48]. Thus, a state may attempt to "upgrade" from mass tourism to heritage tourism in order to enhance its symbolic status, which in turn will generate economic benefits [47].

From this perspective, our findings with regard to regional variation in the perceptions of touristic threats or commendations for model tourism management can be explained by the fact that countries have unequal capacities to exploit their cultural wealth for status advantages. As inscription on the World Heritage List becomes an increasingly salient indicator of a country's cultural wealth, we see clear advantages emerge for European and North American states (*i.e.*, they are more often commended for sustainable management of tourism and their sites are less likely to be seen as threatened by tourism). Ultimately, this enhances the symbolic value of their inscribed sites, thereby reproducing their advantages in the world system. Our approach, which focuses on the assessments of value made by influential intermediaries (*i.e.*, experts at ICOMOS and IUCN), highlights a useful addition to the perspective. To date, cultural wealth scholars have focused primarily on "impression management" undertaken by states to advance their symbolic status but have given little direct attention to how these claims are received by their intended audiences. In this paper, we focus on how claims to symbolic value are evaluated by advisory bodies that act as "gatekeepers" to cultural wealth. Future work in the cultural wealth perspective could draw on sociological scholarship on valuation and evaluation to better understand how claims to cultural wealth are legitimated by various audiences [89]. In the case of world heritage, ICOMOS and IUCN play a central role in articulating and institutionalizing concerns about sustainability and tourism as well as assessing whether such concerns add to or detract from the outstanding universal value of cultural and natural sites.

ACKNOWLEDGMENTS

We would like to thank the J. Murrey Atkins Library at the University of North Carolina, Charlotte for providing financial assistance through the Open Access Article Publication Subvention Fund (OAAPSF) to help cover the costs of publishing in open access.

AUTHOR CONTRIBUTIONS

Both authors contributed to conceptualizing and carrying out the present study as well as writing the manuscript. Both authors read and approved the final manuscript.

CONFLICTS OF INTEREST

The authors declare no conflict of interest.

REFERENCES

1. Convention Concerning the Protection of the World Cultural and Natural Heritage. Available online: http://whc.unesco.org/en/conventiontext/ (accessed on 25 January 2016).

2. Elliott, M.A.; Schmutz, V. World heritage: Constructing a universal cultural order. *Poetics***2012**, *40*, 256–277.

3. States Parties Ratification Status. Available online: http://whc.unesco.org/en/statesparties/(accessed on 25 January 2016).

4. Schmutz, V.; Elliott, M.A. World heritage and the scientific consecration of "outstanding universal value". *Int. J. Comp. Sociol.* **2016**. under review.

5. Barthel-Bouchier, D. *Cultural Heritage and the Challenge of Sustainability*; Left Coast Press: Walnut Creek, CA, USA, 2013.

6. McKercher, B. Some fundamental truths about tourism: Understanding tourism's social and environmental impacts. *J. Sustain. Tour.* **1993**, *1*, 6–16.

7. Porananond, P.; King, V.T. *Rethinking Asian Tourism: Culture, Encounters and Local Response*; Cambridge Scholars Publishing: Newcastle upon Tyne, UK, 2014.

8. Winter, T. Tomb raiding Angkor: A clash of cultures. *Indone. Malay World* **2003**, *31*, 58–69.

9. Labadi, S. *UNESCO, Cultural Heritage, and Outstanding Universal Value*; AltaMira Press: Lanham, MD, USA, 2013.

10. World Heritage and Sustainable Tourism Programme. Available online: http://whc.unesco.org/en/tourism/ (accessed on 25 January 2016).

11. World Heritage and Sustainable Development. Available online: http://whc.unesco.org/en/sustainabledevelopment/ (accessed on 25 January 2016).

12. Evans, G. World heritage and the World Bank: Culture and sustainable development? *Tour. Recreat. Res.* **2001**, *26*, 81–84.

13. Lansing, P.; de Vries, P. Sustainable tourism: Ethical alternative or marketing ploy? *J. Bus. Ethics* **2007**, *72*, 77–85.

14. Buckley, R. Sustainable tourism: Research and reality. *Ann. Tour. Res.* **2012**, *39*, 528–546.

15. Liu, Z. Sustainable tourism development: A critique. *J. Sustain. Tour.* **2003**, *11*, 459–475.

16. Meyer, J.W. World society, institutional theories, and the actor. *Ann. Rev. Sociol.* **2010**, *36*, 1–20.

17. Bandelj, N.; Wherry, F.F. *The Cultural Wealth of Nations*; Stanford University Press: Palo Alto, CA, USA, 2011.

18. Tunbridge, J.E.; Ashworth, G.J. *Dissonant Heritage: The Management of the Past as a Resource in Conflict*; Wiley: London, UK, 1996.

19. Maffi, I. The emergence of cultural heritage in Jordan: The itinerary of a colonial invention. *J. Soc. Archaeol.* **2009**, *9*, 5–34.

20. Peleggi, M. National heritage and global tourism in Thailand. *Ann. Tour. Res.* **1996**, *23*, 432–448.

21. Byrne, D. Western hegemony in archaeological heritage management. *Hist. Anthropol.* **1991**,*5*, 269–276.

22. Cleere, H. The World Heritage Convention in the third world. In *Conservation and Management of Archaeological Sites*; McManamon, F., Hatton, A., Eds.; Routledge: London, UK, 2000; pp. 99–106.

23. Labadi, S. A review of the Global Strategy for a Balanced, Representative and Credible World Heritage List 1994–2004. *Conserv. Manag. Archaeol. Sites* **2005**, *7*, 89–102.

24. Barthel-Bouchier, D. Authenticity and identity: Theme-parking the Amanas. *Int. Sociol.* **2001**,*16*, 221–239.

25. Lowenthal, D. *The Heritage Crusade and the Spoils of History*; Free Press: New York, NY, USA, 1996.

26. Lowenthal, D. *The Past Is a Foreign Country*; Cambridge University Press: Cambridge, UK, 2005.

27. Smith, M. A critical evaluation of the global accolade: The significance of World Heritage Site status for Maritime Greenwich. *Int. J. Herit.Stud.* **2002**, *8*, 137–151.

28. Huang, C.; Tsaur, J.; Yang, C. Does world heritage really induce more tourists? Evidence from Macau. *Tour. Manag.* **2012**, *33*, 1450–1457.

29. Landorf, C. Managing for sustainable tourism: A review of six cultural World Heritage Sites.*J. Sustain. Tour.* **2009**, *17*, 53–70.

30. Adams, K.M. The politics of heritage in Tana Toraja, Indonesia: Interplaying the local and the global. *Indones. Malay World* **2003**, *31*, 91–107.

31. Bandyopadhyay, R.; Morais, D.B.; Chick, G. Religion and identity in India's heritage tourism. *Ann. Tour. Res.* **2008**, *35*, 790–808.

32. Adams, K.M. *Art as Politics: Re-Crafting Identities, Tourism, and Power in Tana Toraja, Indonesia*; University of Hawaii Press: Honolulu, HI, USA, 2006.

33. Chang, T.C.; Milne, S.; Fallon, D.; Pohlmann, C. Urban heritage tourism: The global-local nexus. *Ann. Tour. Res.* **1996**, *23*, 284–305.

34. Jamal, T.; Stronza, A. Collaboration theory and tourism practice in protected areas: Stakeholders, structuring and sustainability. *J. Sustain. Tour.* **2009**, *17*, 169–189.

35. Meyer, J.W. The world polity and the authority of the nation-state. In *Studies of the Modern World-System*; Bergesen, A., Ed.; Academic Press: New York, NY, USA, 1980; pp. 109–137.

36. Boli, J.; Thomas, G.M. World culture in the world polity. *Am. Sociol. Rev.* **1997**, *62*, 171–190.

37. Boli, J.; Thomas, G.M. *Constructing World Culture*; Stanford University Press: Palo Alto, CA, USA, 1999.

38. Frank, D.J. Science, nature, and the globalization of the environment, 1870–1990. *Soc. Forces***1997**, *76*, 409–437.

39. Frank, D.J.; Hironaka, A.; Schofer, E. The nation-state and the natural environment over the twentieth century. *Am. Sociol. Rev.* **2000**, *65*, 96–116.

40. Shandra, J.M.; Esparza, L.E.; London, B. Non-governmental organizations, democracy, and deforestation: A cross-national analysis. *Soc. Nat. Resour.* **2012**, *25*, 251–269.

41. Delgado-Torres, A.; Palomeque, F.L. The growth and spread of the concept of sustainable tourism: The contribution of institutional initiatives to tourism policy. *Tour. Manag. Perspect.***2012**, *4*, 1–10.

42. Bourdieu, P. *Distinction: A Social Critique of the Judgement of Taste*; Harvard University Press: Cambridge, MA, USA, 1984.

43. Reyes, V. The production of cultural and natural wealth: An examination of world heritage sites. *Poetics* **2014**, *44*, 42–63.

44. Centeno, M.A.; Bandelj, N.; Wherry, F.F. The political economy of cultural wealth. In *The Cultural Wealth of Nations*; Bandelj, N., Wherry, F.F., Eds.; Stanford University Press: Palo Alto, CA, USA, 2011; pp. 23–46.

45. Kowalski, A. When cultural capitalization became global practice: The 1972 World Heritage Convention. In *The Cultural Wealth of*

Nations; Bandelj, N., Wherry, F.F., Eds.; Stanford University Press: Palo Alto, CA, USA, 2011; pp. 73–89.

46. Regnault, M. Converting (or not) cultural wealth into tourism profits: Case studies of Reunion Island and Mayotte. In *The Cultural Wealth of Nations*; Bandelj, N., Wherry, F.F., Eds.; Stanford University Press: Palo Alto, CA, USA, 2011; pp. 177–196.

47. Bair, J. Constructing scarcity, creating value: Marking the Mundo Maya. In *The Cultural Wealth of Nations*; Bandelj, N., Wherry, F.F., Eds.; Stanford University Press: Palo Alto, CA, USA, 2011; pp. 23–46.

48. Rivera, L.A. Impression management of stigmatized nations: The case of Croatia. In *The Cultural Wealth of Nations*; Bandelj, N., Wherry, F.F., Eds.; Stanford University Press: Palo Alto, CA, USA, 2011; pp. 114–138.

49. Bourdieu, P. *The Field of Cultural Production*; Harvard University Press: Cambridge, MA, USA, 1993.

50. Tentative Lists. Available online: http://whc.unesco.org/en/tentativelists/ (accessed on 26 January 2016).

51. UNESCO. Operational Guidelines for the Implementation of the World Heritage Convention. Available online: http://whc.unesco.org/en/guidelines/ (accessed on 11 March 2013).

52. Introducing ICOMOS. Available online: http://www.icomos.org/en/about-icomos/mission-and-vision/mission-and-vision (accessed on 6 March 2013).

53. About IUCN. Available online: http://www.iucn.org/about/ (accessed on 25 January 2016).

54. Meskell, L. States of conservation: Protection, politics, and pacting within UNESCO's World Heritage Committee. *Anthropol. Q.* **2014**, *87*, 217–244.

55. Meskell, L. UNESCO's World Heritage Convention at 40: Challenging the economic and political order of international heritage conservation. *Curr. Anthropol.* **2013**, *54*, 483–494.

56. World Heritage List. Available online: http://whc.unesco.org/en/list/ (accessed on 25 January 2016).

57. ICOMOS. *Evaluation of Group of Monuments at Hampi*; ICOMOS: Paris, France, 1986.

58. IUCN. *Evaluation of Iguazu National Park*; IUCN: Gland, Switzerland, 1984.

59. ICOMOS. *Evaluation of the Medieval City of Rhodes*; ICOMOS: Paris, France, 1988.

60. IUCN. *Evaluation of Great Smoky Mountains National Park*; IUCN: Gland, Switzerland, 1983.

61. ICOMOS. *Evaluation of Temple of Apollo Epicurius, Bassae*; ICOMOS: Paris, France, 1986.

62. ICOMOS. *Evaluation of Great Zimbabwe National Monument*; ICOMOS: Paris, France, 1986.

63. ICOMOS. *Evaluation of Chan Chan Archaeological Zone*; ICOMOS: Paris, France, 1986.

64. IUCN. *Evaluation of Kilimanjaro National Park*; IUCN: Gland, Switzerland, 1987.

65. ICOMOS. *Evaluation of Paphos*; ICOMOS: Paris, France, 1980.

66. IUCN. *Evaluation of Royal Chitwan National Park*; IUCN: Gland, Switzerland, 1984.

67. ICOMOS. *Evaluation of Ksar of Ait-Ben-Haddou*; ICOMOS: Paris, France, 1987.

68. ICOMOS. *Evaluation of Tipasa*; ICOMOS: Paris, France, 1982.

69. ICOMOS. *Evaluation of Historic Centre of Evora*; ICOMOS: Paris, France, 1986.

70. IUCN. *Evaluation of Comoe National Park*; IUCN: Gland, Switzerland, 1983.

71. IUCN. *Evaluation of Gros Morne National Park*; IUCN: Gland, Switzerland, 1987.

72. ICOMOS. *Evaluation of Bisotun*; ICOMOS: Paris, France, 2006.

73. ICOMOS. *Evaluation of the Red Fort Complex*; ICOMOS: Paris, France, 2007.

74. IUCN. *Evaluation of Socotra Archipelago*; IUCN: Gland, Switzerland, 2008.

75. ICOMOS. *Evaluation of Tabriz Historic Bazaar*; ICOMOS: Paris, France, 2010.

76. ICOMOS. *Evaluation of Sulaiman-Too Sacred Mountain*; ICOMOS: Paris, France, 2009.

77. ICOMOS. *Evaluation of the Sacred Mijikenda Kaya Forests*; ICOMOS: Paris, France, 2008.

78. IUCN. *Evaluation of Monarch Butterfly Biosphere Reserve*; IUCN: Gland, Switzerland, 2008.

79. ICOMOS. *Evaluation of Fujian Tolou*; ICOMOS: Paris, France, 2008.

80. ICOMOS. *Evaluation of Old Town of Regensburg*; ICOMOS: Paris, France, 2006.

81. ICOMOS. *Evaluation of Genoa: Le Strade Nuove and the System of the Palazzi dei Rolli*; ICOMOS: Paris, France, 2006.

82. IUCN. *Evaluation of Jeju Volcanic Island and Lava Tubes*; IUCN: Gland, Switzerland, 2007.

83. ICOMOS. *Evaluation of Episcopal City of Albi*; ICOMOS: Paris, France, 2010.

84. ICOMOS. *Evaluation of Seventeenth-Century Canal Ring Area of Amsterdam*; ICOMOS: Paris, France, 2010.

85. Gullino, P.; Becarro, G.L.; Larcher, F. Assessing and monitoring the sustainability in rural world heritage sites. *Sustainability* **2015**, *7*, 14186–14210.

86. Elliott, M.A. The institutionalization of human rights and its discontents. *Cult. Sociol.* **2014**, *8*, 407–425.

87. Bieri, F.; Boli, J. Trading diamonds responsibly: Institutional explanations for corporate social responsibility. *Sociol. Forum* **2011**, *26*, 501–526.

88. Jang, Y.S.; Cho, M.; Drori, G.S. National transparency. *Int. J. Comp. Sociol.* **2014**, *55*, 95–118.

89. Lamont, M. Toward a comparative sociology of valuation and evaluation. *Ann. Rev. Sociol.***2012**, *38*, 201–221.

7

CHAPTER

There Is No Such Thing as Sustainable Tourism: Re-Conceptualizing Tourism as a Tool for Sustainability

Gianna Moscardo † and Laurie Murphy †

School of Business, James Cook University, Townsville QLD 4811, Australia

ABSTRACT

Increased global concern about sustainability has placed pressure on businesses to justify the value of their products and services beyond personal profit and to take responsibility for the negative impacts of their activities. Tourism is particularly susceptible to this pressure, given its generally poor track record in terms of negative social, cultural and environmental impacts, and the lack of compelling evidence of benefits for either the individual tourist or destination communities. While the management of tourism impacts and the relationship between tourism and sustainability have been paid considerable attention by tourism academics, there is little evidence of any significant change in tourism practice. This paper will argue that this lack of change reflects problems in the way tourism academics have conceptualized sustainable tourism. After reviewing these problems with sustainable tourism, this paper will offer an alternative framework for sustainable tourism that focuses on the concept of quality-of-life, recognizes the complexity of tourism within local and global systems, adheres to the principles of responsible tourism, and explicitly assesses the value of tourism as one tool, amongst many, for sustainability. One potential application of the framework will be demonstrated with a case study of tourism development on Magnetic Island in Australia.

Keywords: tourism sustainability; quality of life; sustainable tourism; tourism planning

1. INTRODUCTION

A recent international media headline read "Hawaiian town of Kailua asks state to stop recommending it as a tourist destination" [1]. According to the article, the Neighborhood Board lobbied the state tourism agency to stop promoting tourism to their community reporting significant negative impacts of tourism that far outweighed any benefits. These negative impacts included inflation, increases in housing prices, loss of social networks, an influx of temporary strangers, increased crime, and tourists demonstrating undesirable behavior to local children [1]. The story seems sadly familiar, both in Hawaii, where Cohen [2] describes a 1971 media story with the headline "Please don't visit Hawaii until we can save what's left" (p. 227), and in many other destinations. While, the style of tourism may have changed over the forty years between these stories, the fundamental problem of trying to balance the competing pressures and impacts associated with tourism remains. Despite more than 30 years of academic attention and much discussion of sustainable tourism (ST), it appears that little has changed in the practice of tourism in many places. This paper will review some of the problems identified with the concept of ST. In particular it will argue that we need to very clearly distinguish between the concept of ST and the idea of tourism as one possible tool to support sustainability at multiple levels. Taking the idea of tourism as a tool for sustainability the paper will further suggest that a more explicit consideration of how tourism affects all aspects of quality of life could support more positive tourism contributions to sustainability. The paper will propose a new framework for tourism analysis and planning and demonstrate its potential application in a case study of tourism on Magnetic Island, Australia.

1.1 Tourism and Sustainability: Confusion and Conflict

In Jafari [3] proposed four tourism research and policy platforms that described different approaches to analyzing and planning tourism. The first was the Advocacy platform, referring to the presentation of tourism as an effective economic development strategy providing employment and income. The second was the Cautionary platform, which highlighted the negative impacts of tourism, arguing against tourism because of its potential as a destroyer of societies and environments. The third was the Adaptancy platform, which argued that it was possible to have the economic benefits without the social and environmental costs if alternative or adapted forms of tourism could be implemented. Jafari used the platforms as a guide for the development of tourism education and research programs arguing that each of the first three platforms was limited and that tourism academics need to take a more scientific or Knowledge Based platform [3]. Although these four platforms emerged sequentially over time, all continue to exist and it is in the interplay between the first three platforms that we can see the beginning of concerns about tourism and sustainability. Indeed Macbeth in 2005 [4] argued that tourism and sustainability could be seen as a fifth platform given the considerable attention that tourism academics have paid to this relationship since the early 1990s.

One major outcome of this attention has been the rise of the concept of ST. Several authors have provided historical reviews of this concept and its previous and current use in tourism [5,6,7]. What is of most relevance to this special issue and the present paper are the critiques of its formulation. Five main issues with regard to ST have been identified—it is tourism or tourist-centric, it is focused on the destination space and ignores the larger global system that tourism is a part of, it highlights environmental concerns and often ignores other dimensions of sustainability, it is usually driven by academics and institutions that present sustainability as an objective, scientific concept best achieved by the development and enforcement of rules and guidelines for specific practice, and it is not integrated into other activities [5,6,7,8,9,10].

Moscardo [11] repeated Getz's [12] review of tourism planning models and concluded that very little had changed in thirty years in the way tourism planning processes are conceptualized and used in academic and government documents. Both reviews identified problems that included:

- A narrow focus on specific projects, rather than considerations of tourism as a whole;
- Limited attention given to tourism impacts;
- A focus on economic factors with occasional limited acknowledgement of environmental issues;
- A failure to consider how tourism would interact with, and effect other activities at a destination;
- The naïve adoption of business strategic planning as the dominant framework for tourism planning, and, as a consequence;

- The placement of market or tourist needs and expectations as the core drivers of tourism planning, giving destination residents a very limited role, if any.

This placement of tourist or market needs as the core driver of the process, and the often unstated and unchallenged assumption that a viable and profitable tourism industry is the end goal of tourism planning has unfortunately also been adopted in most applications of ST [7,10,13,14]. Guidelines for ST practice assume that some form of tourism practice is sustainable and rarely, if ever, is "no tourism" considered as a serious option.

ST is typically translated into sets of guidelines and rules for practice that focus almost exclusively on immediate, local, environmental impacts and are driven by academics and organizations that adhere to a view of sustainability as an objective concept supported by science [4,5,6,7,9,10,13,14]. While it is important to address immediate local environmental impacts of tourism, a failure to consider the wider, cumulative, long-term and other impacts of tourism means that it may be possible to label an activity as ST even if it fails to address the bulk of the issues identified in the wider sustainability literature. A critical examination of the application of ST to the Great Barrier Reef provides an example of this problem. While climate change is identified as a major threat to this natural environment, a distinguishing characteristic of many tourists to the destination is that they have to take long haul flights or drive considerable distances to get the destination, thus contributing directly to the problem of climate change. Discussions of ST in this destination do not, however, include consideration of the carbon emissions associated with travel to the destination [15]. This problem is not just limited to spatial and temporal limitations. According to Macbeth [4] ST is often locked into a conservation ethic and so emphasizes environmental issues adopting an ecological approach that "is susceptible to social indifference" (p. 968). These gaps in ST are partly the result of the fourth problem; the dominance of external experts in the development of ST rules, guidelines and standards which assume a shared objective scientific view of sustainability. This framing of ST further disempowers other stakeholders, especially local residents [7,8,9,10], ignores the value-laden, subjective, cultural nature of sustainability, and thus fails to address the justice issues that are critical to understanding the origins of unsustainability [4,6,7].

Jafari provided an update to his four original platforms and argued for a fifth or Public platform in which tourism would look outwards towards other activities and other sources of knowledge [15]. In this argument he described tourism as "self-endorsing, inward-looking [and] narcissistic" (p. 2). While Jafari was referring to tourism research and practice as a whole, ST is particularly problematic in this regard with no consideration given to the extent to which tourism in any place or at any level effects other socio-economic development strategies. Tourism is assumed to be not just an end in itself, but a solitary activity that neither impacts on, nor is impacted by, other sectors. This failure to consider integration with other development actions also limits the extent which tourism can be seen as sustainable [12,16].

Suggestions for addressing these issues fall into four categories. The first and most common is a more explicit consideration of ethical approaches to tourism and issues of social justice [4,7,17,18]. These discussions direct tourism researchers to consider more carefully their own personal ethics and the impacts of their actions on others [4]. According to Jamal, Camargo and Wilson [7], in their contribution to this special issue, the use of ethics and justice to guide tourism towards sustainability is better served by the idea of responsible tourism rather than ST. Responsible tourism is seen as tourism that:

- Involves the host community in tourism governance;
- Actively seeks to minimize all negative impacts;
- Generates economic benefits for locals, contributes to cultural and natural heritage, and enhances the overall well-being of the host community;
- Is accessible to all; and
- Provides tourists with meaningful experiences that give them a better understanding of cultural, social and environmental issues relevant to the places they visit as well as the larger, global setting [19].

Responsible tourism is consistent with the second category of suggested future directions, which is the explicit recognition and assessment of tourism in terms of well-being or quality of life (QoL). These two terms, well-being and QoL, are often used interchangeably. The OECD [20], for example, defines QoL as "the notion of human welfare (wellbeing) measured by social indicators rather than by 'quantitative' measures of income and production" (p. 1). QoL refers to the idea that a good and satisfying life depends on meeting key basic needs [21,22]. These needs are physiological, including food, water, good health, and security, opportunities for education, productive activity and involvement in relevant decisions, belongingness, and self-esteem. These needs are, in turn, met through access to a set of capitals described in Table 1 [21,22,23]. Well-being is often seen as the broader term and so for the purposes of this paper the term QoL will be used for the individual level of analysis, while the term community well-being (CW) will be used at the social level of analysis. This differentiation is provided to make subsequent discussion easier for the reader to follow.

Definitions of sustainability increasingly use the idea of well-being or QoL and the importance of multiple forms of capital as central elements, suggesting that sustainability is about increasing all forms of capital, not just financial or built capital, recognizing that natural capital is especially important because it cannot be substituted with other forms [24,25,26]. The importance of multiple forms of capitals and QoL has been recognized in recent tourism and sustainability discussions [9,21,27,28,29,30] with a growing number of researchers arguing for the use of QoL as way to assess the sustainability of tourism [9,21,27,28,30].

Table 1. Main Types of Capital that Support QoL. Source: [24] (p. 534).

Type of Capital	Description
Financial	Income, savings, and access to funding for investment.
Natural	Natural ecosystems and the assets, services and resources that they provide includes landscape, environmental systems, green spaces and conservation areas
Built	Physical facilities and infrastructure that communities have available for use including buildings, transport systems, public spaces, technological systems and distribution systems for water, waste and energy.
Social	Features of social networks such as trust, reciprocity and cooperation and social institutions and associations
Cultural	Values and symbols shared by human groups and manifested in things such ritual and social activities, arts and crafts, spiritual practices, languages and celebrations
Human	The capabilities, skills, knowledge and health of the people who make up a community
Political	Ability to access political decision making processes and influence governance

Moscardo, for example, proposes a QoL framework for considering tourism impacts that suggests that different types of tourist activity can be assessed according to their positive and negative impacts on each of the capitals listed in Table 1 [21]. Of importance to the present discussion is that this framework also argues that the assessment needs to be done at several levels including the individual tourist, stakeholders in the regions that tourists leave, stakeholders in transit regions, stakeholders in destination regions, and the people who work in tourism [21]. According to this framework ST has to result in net benefits across all the types of capital for the individual tourists, the generating, transit and destination communities, the people who work in tourism and more globally. This recognition that tourism exists simultaneously at multiple levels is the third theme in suggested new directions for ST [7]. The argument here is that assessment of tourism impacts must not be limited to the destination level. The growing literature on the contributions of tourism to climate change [31] is a clear example of how tourism impacts extend beyond the destination. In a similar fashion it could be argued that tourism that exploits the global North–South divide by offering experiences in the global South based on the lower costs resulting from lower standards of living is also not sustainable in a global sense [7,32].

Finally, there are calls to devise new approaches to tourism planning. Moscardo's 2011 [11] review of tourism planning models and approaches identified a consistent set of patterns in terms of the number and sequence of steps or phases in the planning process, the identification and engagement of key stakeholders and the underlying assumptions. The main planning steps in their most common sequence are presented in Figure 1. Traditional formal approaches to tourism planning are typically focused on the destination level and generated by groups external to the destination community such as regional or national government agencies or development organizations. These external agents then usually appoint a planning group of some sort who take responsibility for the plan and set the objectives or goals. Within this traditional process it is always assumed that some form of tourism is desirable and the goal of the plan is to ensure that tourism as a commercial activity is financially successful for tourism businesses, regardless of who owns those businesses. The second common step is to analyze the situation combining both an assessment of the destination and its residents in terms of the assets and resources they can provide to support a viable tourism sector and an examination of potential tourist markets. This information is then used to develop plans or strategies and following that to move to implementation and monitoring. It is typically presented and considered as a linear process. Where opportunities for community or public engagement or participation are provided, and that is not always the case, they are most often included after the development of the plans/strategies and focused almost exclusively on generating public support for the decisions already made. In some cases there is an opportunity for some input from the destination community into the development of the plans themselves. Other critical reviews of tourism planning processes and development policies by Hall [33] and Ruhanen [34] provide consistent conclusions to those described by Moscardo [8,11]. These critiques further argue that, despite the addition of sustainable tourism as a goal in many plans and increased attention paid to environmental impacts of tourism, this traditional approach to tourism planning remains largely unchanged and thus is unlikely to improve the sustainability of tourism.

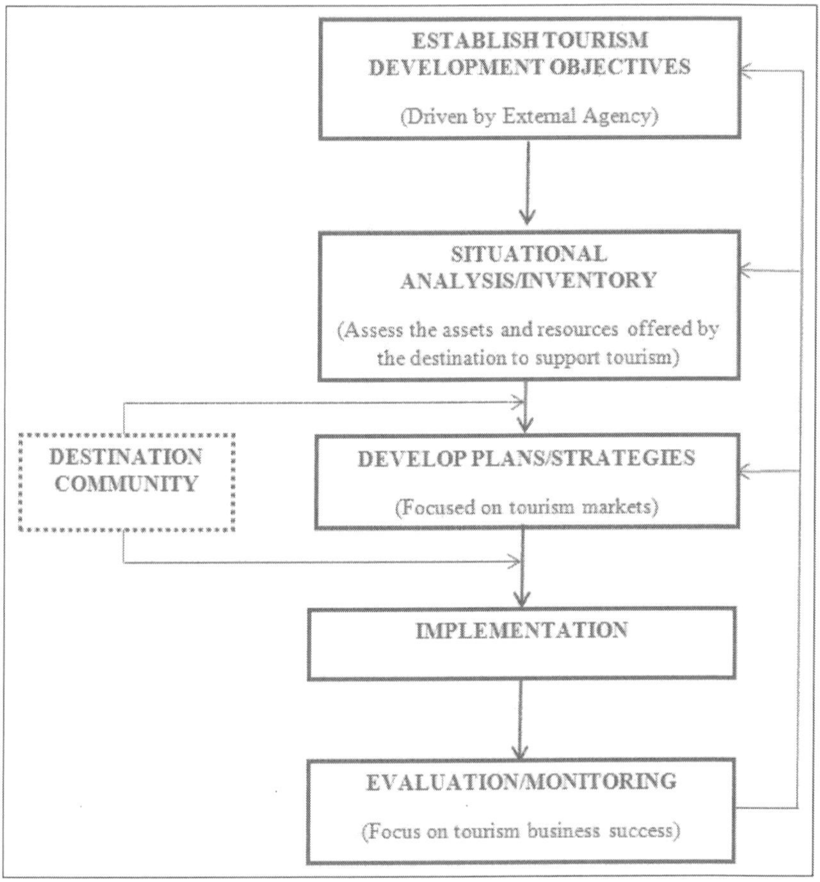

Figure 1. The traditional tourism planning process.

The analysis thus far suggests that new approaches to tourism planning and development are needed. These new approaches should take into account the need for multiple levels of analysis and address QoL beyond simple economic and environmental dimensions [4,11]. New approaches to tourism planning would also need to respond to the challenges of effective governance and public participation, giving a stronger voice to destination residents and involving a wider range of stakeholders than those directly involved in tourism [5,11]. If tourism is to contribute to sustainability at all levels then arguably tourism planning should also involve some form of sustainability evaluation of tourism development proposals before they are implemented. Despite a growing literature on sustainability assessment of tourism, the proposed systems are designed for existing tourism activities and the outcomes of tourism developments

[35,36]. Such a pre-implementation assessment would more directly consider tourism as one possible tool for achieving sustainability and examine its value as a means to an end, rather than an end itself. Thus proposed tourism projects would be assessed according to the resources that they would provide for use by residents, the assets that would be enhanced and the opportunities that would be provided for other commercial, social and cultural activities [8]. Rather than asking what resources are available in the destination for tourism, this type of sustainability assessment would ask what resources tourism would bring to the region and how tourists could be used to meet the needs and aspirations of destination residents and non-tourism businesses.

In summary the problems identified with current conceptualizations of ST and traditional tourism planning approaches stem from three unquestioned assumptions:

- That tourism is an end in itself and so tourists are the key stakeholders of interest;
- That tourism is primarily a source of financial capital which is assumed can be exchanged for other capitals, and
- That negative tourism impacts are mostly confined to the physical environment of the specific destination.

If tourism is to move towards a greater contribution to sustainability there is a need to reconsider the relationships between tourism and sustainability and to find, apply and test new planning approaches. The present paper seeks to address these needs by outlining a re-conceptualization of tourism development and sustainability that replaces these common assumptions and explicitly casts tourism as a means to improving well-being, rather than an end in itself, that the key stakeholders are those that have to live with consequences of tourism, and that tourism impacts extend beyond the destination in multiple dimensions and across multiple levels of analysis. As such the paper is not filling a research gap but rather seeks to address these conceptual issues in the dominant planning paradigms for tourism. To address this need it describes a possible alternative tourism planning framework and then uses a single case study to critically analyze how this new framework might work in practice.

2. RE-CONCEPTUALIZING TOURISM AS A TOOL FOR SUSTAINABILITY

This section outlines an alternative approach to conceptualizing the relationship between tourism and sustainability using a Quality of Life (QoL) approach. This QoL approach is used to build an alternative tourism planning framework for destinations that seeks to address the issues identified in the previous review sections. Figure 2 provides an overview of a QoL approach for assessing the sustainability of tourism in general. This approach proposes that in order to be considered sustainable, the tourism development or business must:

- Demonstrate net positive contributions to each of the forms of capital listed on the right hand side of the diagram, paying particular attention to natural capital;
- Be applied at all levels; and
- Consider all the issues listed on the left hand side of diagram.

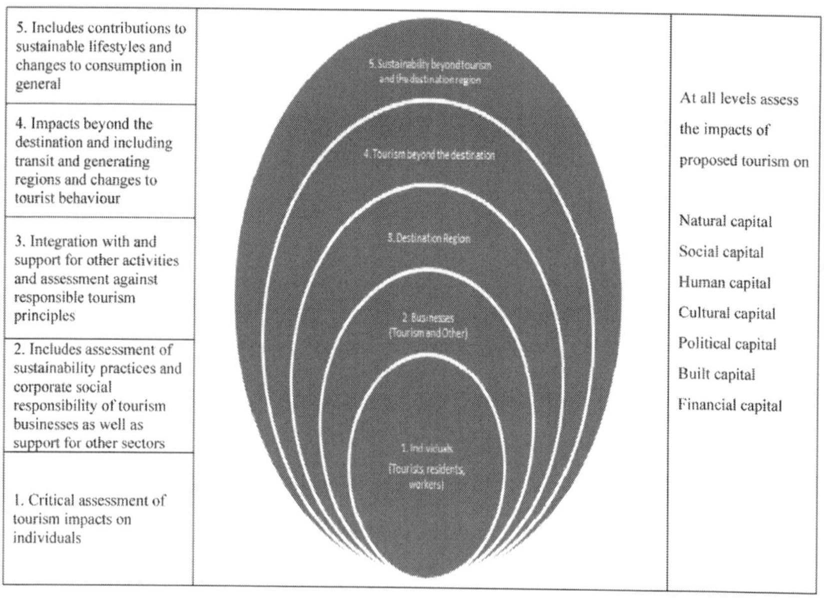

Figure 2. A quality of life (QoL) approach for assessing tourism sustainability.

For example, a proposal to develop farmstays in a rural region would have to demonstrate the benefits of the activity for all the forms of capital for the host farmers and their families, other residents and business in the destination region, and the tourists themselves. These analyses would include a consideration of any opportunity costs associated with the proposal and examine how the activity would contribute to other activities. The proposal would have to detail the risks and management responses associated with negative impacts on all the forms of capital for all the levels. At the level of the business the model is similar to the concept of sustainable livelihoods [37]. The approach goes beyond this to include other businesses outside of tourism, the working conditions of tourism staff, the benefits and costs for the tourists, and the impacts of the tourism beyond the destination including such things as the carbon emissions for the associated travel to the region and impacts on transit regions. Finally the assessment of this proposed tourism activity would have to consider how it influences sustainability beyond tourism. It could be argued, for example, that the presence of tourists might support shifts towards more sustainable agricultural practice on the farms and that exposure to the daily lives of farmers, supported by appropriate interpretation, might encourage tourists to consider and change their subsequent food purchases.

This example shows how the QoL approach could be used to assess tourism proposals and through that assessment how proposals could be revised to improve contributions to sustainability. The approach can be used to guide tourism planning at multiple levels. At the individual level with a focus on tourists, for example, it could be used to support more detailed and effective guidelines for being a responsible tourist. At the level of sustainability beyond tourism it directs attention to the potential role of tourism as a form of social marketing to support sustainability action [38]. At the destination level it provides a way to compare different types or forms of tourism and to compare tourism to other socio-economic development options.

The QoL approach to assessing the sustainability of tourism outlined in Figure 2 can also be used to support alternative approaches to tourism planning. Although tourism planning can be conducted at a number of levels, regional or destination tourism planning remains the dominant point at which tourism proposals and projects are evaluated and the framework can be used to suggest alterations to traditional tourism planning processes at this level. Figure 3 provides a basic outline of an alternative framework for destination tourism planning based on the QoL approach. As this particular framework is applied to the destination it is focused on Community Wellbeing (CW).

The framework places the destination community at the centre of tourism planning activity, strengthening community engagement across the whole process. A common conclusion in the literature on public participation in development planning is the need to move from simply informing residents about the plans or seeking their input into plans established by others, to empowering residents to control the planning and development process [39]. Thus a critical element of tourism planning has to be finding ways to build community capacity to effectively make decisions about tourism. Moscardo [21] addresses in more detail the dimensions of tourism governance needed to achieve this, highlighting the importance of developing effective leaders, networks and partnerships and, in particular, educating destination residents and stakeholders about tourism, especially its impacts [40]. The development of effective governance for tourism planning is not likely to be easy in practice and will vary according to the nature of the destination community and the larger political, social, and cultural context. Marzuki and Hay [39] offer a system that could be adapted to a wide range of situations to assist in the development of such governance structures.

Reviews of community development beyond tourism also challenge the role of external consultants arguing for a shift away from consultant experts to facilitators and coordinators with longer term relationships with the communities in question [41]. Feighery makes a similar argument in his discussion of the ethics of academics acting as tourism development consultants, outlining in detail the conflicts of interest that often accompany the work of tourism academics [42]. While the framework described in Figure 3 does not include a specific element it is recognized that external agents will inevitably be involved in tourism planning. The framework suggests that such external agents should be assessed in terms of the resources they offer to the tourism planning process.

As destination communities develop their capacity for tourism governance, the CW framework re-conceptualizes tourism as a tool for destination stakeholders and sets the main goal of tourism planning as supporting improvements in wellbeing for destination communities and achieving improvements to sustainability at a number of levels. Instead of assessing the resources available for tourism, this approach argues for an assessment of the stock of the various capitals available to destination residents, and the major sustainability issues that face the destination. The aim being to determine the destination needs rather than tourist attraction potential. Once these needs are identified, the process moves to the generation of tourism options that might address these needs. This requires the use of futures scenario building techniques that are not often used in tourism planning [43].

Once the range of possible tourism scenarios is identified, the next stage in the planning process assesses each scenario according to the resources it offers to the destination, its likely impacts, both positive and negative, on the different forms of capital that make up CW, its business viability and its sustainability with regard to the other four levels of the QoL assessment approach. The viability assessment is especially important as many alternative approaches to tourism such as community based tourism often fail to consider the market support for, and competitiveness of, tourism projects [44]. The sustainability assessment at this stage should also look at issues beyond the destination including costs and benefits for the tourists, the nature of the working environment for tourism staff, impacts on transit and generating regions and contributions to sustainability beyond tourism. The last two stages, implementation and monitoring are similar to those put forward for traditional tourism planning approaches.

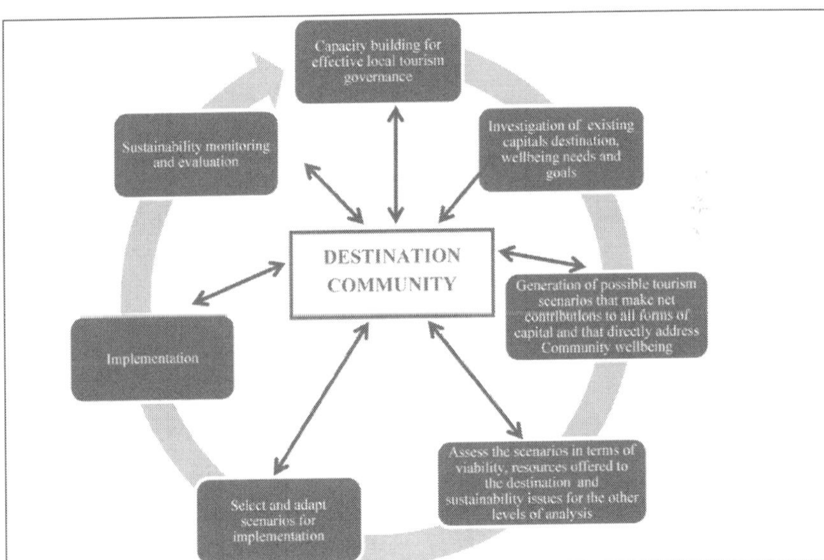

Figure 3. A community well-being approach to destination tourism planning.

3. RE-CONCEPTUALIZING TOURISM AS A TOOL FOR SUSTAINABILITY IN PRACTICE: AN APPLICATION CASE STUDY

This section will describe the potential application of these proposed alternative approaches to tourism and sustainability using a case study of tourism development and planning on Magnetic Island in Australia. This location was chosen as a case study for three main reasons. Firstly, the history of tourism development in this location is very similar to that described in many other places and these similarities will be noted where relevant in the case study description. This history of tourism development also highlights many of the issues raised in the previous sections. Secondly, the community in this location has been involved in a number of sustainability initiatives, has begun to question the role of tourism in supporting their broader sustainability aspirations, and has been explicitly seeking alternative styles of tourism and approaches to tourism development. Thirdly, Magnetic Island is located close to the university where the authors are employed making it easier to be involved with the whole process more intensively and extensively and offering an opportunity to be engaged in tourism research and outreach that made a more direct contribution to sustainability in practice. The aim of using a case study is to demonstrate how the proposed planning framework might be applied in practice. The case study is a demonstration or example of application, not a specific research study, so the focus of the case study description is on the overall sequence of events and actions with some key results from different phases to explain the process overall.

Magnetic Island (MI), with an area of 52 km^2 (20 square miles) is located eight kilometers (5 miles) off the northeastern coastline of Australia adjacent to the regional city of Townsville [45]. More than half of MI (65%) is National Park and much of the island is inaccessible to vehicles. It is a suburb of Townsville with a population of approximately 2300 residents, many of whom commute daily by ferry to the mainland for work and to attend school [46]. The island is home to a wide range of wildlife, has a number of historical sites related to both traditional use by local Indigenous groups and European settlement, especially in relation to World War Two forts and observation posts [45]. MI is also part of the Great Barrier Reef World Heritage Area and Marine Park and its beaches and fringing coral reefs attract local recreationists and tourists.

3.1. A Brief history of Tourism Development on Magnetic Island

MI has been a popular recreational site for Europeans since settlement in the area in the 1800s with Townsville residents sailing there for day trips and picnics [47]. From the late 1890s until the early 1980s there was a steady increase in both residential growth and the development of small scale tourist accommodation and facilities [48]. In 1984 the announcement of a major tourist development proposal triggered twenty years of conflict over the nature of tourism development on MI. This development, which is most commonly referred to as the Nelly Bay Harbour project [49], has dominated considerations of tourism on MI to the present time. The original proposal was for an extensive complex including accommodation for more than 1300 tourists, nearly 100 home sites, an all-weather ferry terminal, a marina, tavern, sports complex, commercial center and new beach. The proposal involved significant alterations to the existing landscape including the removal of a headland and major reclamation and dredging [49]. The announcement marked the start of an extended period of permit and approval applications, impact assessments, public objections and court cases, accusations of political bribery, multiple changes in ownership and the financial collapse of several of the business stakeholders and ultimately the development consortium itself in 1998 [49,50,51]. In 2003 a much smaller development consisting of a ferry terminal, one resort and a marina was completed [49]. The development site continues to create problems with significant areas of construction unfinished and in a decaying state, and ongoing environmental issues [52]. Most importantly for the present discussion is the legacy of conflict and mistrust amongst residents [49,50,51,53,54]. According to Moyle, Croy and Weiler [54], while there is recognition of the economic and infrastructure benefits of tourism, many residents continue to be opposed to any form of tourism and this can be expressed in hostility towards tourists. Many residents also report feeling a sense of no control over development decisions and disengagement from community life.

Arguably the development has also had negative impacts on tourism to the island, with widespread media coverage of the environmental damage, losses to tourism businesses who invested in the destination based on the development promises, the presence of the unfinished construction at the main arrival point for visitors, and confusion over the style of experience on offer. This sequence of events and patterns of impacts are consistent with those reported in many other locations. The reader is directed to Asher, Roe and Goodwin [55], Blackstock [56], Dredge [57], Holladay and Powell [58], Lee, Riley and Hamilton [59],

Moscardo [21], and Park Lee, Choi and Yoon [60] for examples of similar histories for a range of different locations and for different types of tourism.

The wider region attracts nearly one million tourists annually, made up of approximately 75% domestic and 25% international tourists, and around one quarter of these tourists visit MI [61]. It is difficult to estimate the number of recreational visits by regional residents, but generally there is a perception that all visitor numbers could be improved and that MI tourism businesses have faced significant challenges in recent years due to the global financial crisis, a strong Australian dollar, and damage from cyclones.

3.2. Recent Tourism Planning Exercises

In this setting formal tourism plans are made by the state government using traditional tourism planning approaches which include consultations with "government agencies, tourism industry, tourism stakeholders, developers and investors" [62] (p. 3). Two planning documents were relevant at the time of the start of the MI project, the Tourism Opportunity Plan (TOP) [63] and the Destination Management Plan (DMP) 2007–2010 [62]. The main goal of the Destination Management Plan was to increase visitor numbers, length of stay and visitor expenditure in the wider region. Secondary aims included improving overall marketing impact and promoting sustainable management and development of tourism. The bulk of the plan focusses on improved marketing to existing groups of tourists and capitalizing on the traditional assets of a tropical climate, beaches and proximity to natural environments. The aim of sustainable management is addressed with strategies aimed at enhancing the ongoing viability of tourism with statements such as "maintaining and expanding low-cost air access" [62] (p. 11). Consideration of sustainability beyond this is restricted to ensuring "ongoing protection of the destination's natural attractions" and "distribution of up-to-date information on the effects of global warming" [62] (p. 11). Discussion of resident and non-tourism stakeholder engagement is covered in the following two statements:

- Continue to promote the value of tourism and economic benefits to the community (including jobs, career paths) to encourage support of holiday, visiting friends and relatives and business visitors;
- Maximize community support for tourism by increasing awareness of how the community's interaction with visitors is central to the quality of the visitor experience [62] (p. 12).

This plan is supplemented by the TOP which lists and describes a set of tourism projects and infrastructure development required to achieve the goals set in the DMP. In this plan MI is identified as a Sustainable Destination reflecting its location on the Great Barrier Reef and the commitment of the residents to a number of sustainability initiatives [63]. These plans and the overall planning approach are very much like those identified and critiqued in reviews of tourism planning approaches [33,34].

3.3. Magnetic Island Sustainability Aspirations

In recent years MI residents and businesses have pursued a number of sustainability initiatives. The most notable project has been involvement in the Solar City program [64]. As part of this program many residents and businesses installed solar panels to generate electricity with one of the highest residential uptakes of solar panels in the state [65]. The Solar City program also involved household energy audits, the establishment of a small education center on energy efficient suburban living, and has won an Australian Sustainable Cities Award [66]. The program has also postponed the need to install an additional sub-marine power cable from the mainland [67]. Water is piped to MI from the mainland and the local government, with the support of the residents, has invested in a water recycling plant and a number of programs to limit both the use of water and disposal of waste water into the ocean [68]. Other initiatives that have been discussed for the suburb include a ban on plastic bags and bottles, a move to electric powered public transport vehicles and the desire to become a carbon neutral suburb. These sustainability initiatives have been supported by MI tourism businesses. In 2010 a community forum was held to seek widespread input on how to build more sustainable tourism on the island. The key conclusion from these discussions was the need to manage tourism to be consistent with community sustainability aspirations and to develop new forms of tourism that could actively support these aspirations. Nineteen of the local tourism businesses have signed up to a ST program requiring them to audit their energy, water and carbon use and waste generation and commit to programs to improve performance in all these areas [69].

3.4. Assessing Tourism as a Tool to Support Sustainable Lifestyles

In 2010 the authors were involved in discussions with different stakeholder groups about the challenges of tourism and sustainability on MI as part of conducting research into tourism in the wider region. A change in the ownership of the passenger ferry that provides the main access to MI also prompted a series of discussions around how to revitalize tourism to MI. In 2011 the state government called for proposals to examine in more detail the feasibility of the projects described in the TOP. This represented an opportunity to work within the existing government planning system to attempt to develop tourism strategies that would be more sustainable and make a positive contribution to CW on MI. The authors were able to organize and coordinate a group consisting of representatives from local government, the regional tourism organization, MI tourism businesses, and MI community organizations to provide matching funding, successfully apply for the grant and guide the project. The funding opportunity was identified by several stakeholders and the decision to pursue it taken at group meetings. The actual proposal was directly funded by the ferry company, Sealink, the local city council and the state government development agency and indirectly supported by the university, the regional tourism promotion and marketing organization, and several MI business and community groups. Thus the main impetus for the project came from within the MI community and tourism businesses.

The proposed project was a planning and action research exercise to determine what, if any, type of tourism or tourist would support the sustainability aspirations and overall well-being of the residents and business operators of MI. The project was seen as a trial of the both framework and the planning approach outlined in Figure 3 with the starting point being local stakeholder needs and aspirations, with a focus on developing community capacity for tourism governance and explicitly assessing tourism as a tool for sustainability. The project was guided by the following questions.

- What are the key dimensions of QoL for MI residents?
- What sustainability initiatives are most important to MI residents and business operators?
- What kinds of tourism development could support the sustainability aspirations of the MI residents and businesses?
- What level of support is there amongst current tourists for these sustainability initiatives?
- What are the characteristics of tourist markets that are most supportive of these sustainability initiatives?

The authors acted as facilitators for the project, organizing and coordinating meetings and stakeholder inputs and liaising between different agencies and groups. They also acted as a resource providing examples from other locations and conducting research required by the stakeholders. The project involved continuous input from representatives of the MI community groups and the relevant government agencies at all stages. It began with the lists of community sustainability aspirations already outlined and developed through a range of community engagement activities and a series of meetings to determine required research and to develop sustainable tourism scenarios for assessment. Then the authors coordinated a survey of tourist and regional residents to assess the viability of these scenarios and general support for MI sustainability initiatives. After the completion of the survey the authors facilitated public meetings to revisit and confirm community well-being needs and to gather community perspectives on the different sustainable tourism scenarios.

3.4.1. Resident and Business Consultation Outcomes

In keeping with the CW destination planning framework the project was focused on the needs and aspirations of MI residents and involved MI stakeholders in all aspects of the project with regular group and individual meetings throughout the project. In the early stages the focus was on clarifying community aspirations for sustainability and developing specific tourism scenarios that might support these aspirations. In the later stages there were open public meetings that used futures visioning exercises to explore the key dimensions of MI community wellbeing, sought input into decisions about the sustainability initiatives and the tourism development options, and presented visitor research results. Two key themes emerged from the discussions about the important features that contribute to CW—the sense of community and that the suburb was embedded in the natural environment. The first theme is best summarized by one of the public meeting attendees who stated that “it is a small, friendly community with a village like atmosphere”. The second theme can be summarized with another quote, “the opportunity to live in harmony with the environment”. When asked about things that were needed to improve community wellbeing on MI, four key themes were identified:

- Improvements to public transport;
- A need to maintain or enhance public services such as education and health which can be difficult with a smaller population in a less accessible location;

- Support to maintain and expand the sustainability activities and programs; and
- A desire to find ways to use tourism as a support for the sustainability activities in particular, and MI wellbeing in general.

The most important sustainability initiatives were the Solar City program, a desire to be a carbon neutral suburb, supporting better management of the National Park, moving to electric powered public transport, encouraging more tourism businesses to adopt specific sustainability initiatives, and an interest in education for visitors about both the natural environment and the sustainable lifestyles. Other programs suggested but given lower priority included eliminating plastic packaging and bottles and water conservation. Consultation with the various stakeholders identified four tourism scenarios that were both consistent with MI sustainability aspirations and responsible tourism principles:

- Electric bicycle touring package;
- Environmentally friendly camping options;
- Guided Indigenous tours; and
- Activities supporting an art/cultural precinct.

3.4.2. Tourist and Mainland Resident Survey Outcomes

In order to analyse the perspectives of both current and potential tourists as well as mainland residents who see MI as an important recreational setting, a survey was conducted in a variety of locations including on the ferry between MI and the mainland, at various sites on the island, at major tourist attractions and transport nodes on the adjacent mainland, and at major mainland resident recreation sites, with an online option also available through various regional web links. The survey was conducted over a three week period covering local school holidays, weekends and weekdays and the major tourist season. The survey was completed by 884 respondents with just over half (51%) surveyed at MI or on the MI ferry. The sample was relatively evenly split between regional residents (49%) and tourists from outside the region (51%) and included both respondents who had visited MI recently (78%) and those who had either never visited or had not been in the previous 12 month period (22%). The sample had a mean age of 36 years (SD = 14.5) and consisted of 36% males and 64% females.

The questionnaire was developed in consultation with key stakeholders and included questions to profile the respondents on key socio-demographic features, to describe their travel experiences on Magnetic Island, and to examine their evaluations of, and responses to, the proposed sustainability initiatives. Initial analyses to determine what, if any, significant differences existed between regional residents and tourists from outside the region found very few differences. There were some differences in travel experiences with MI including regional residents having a greater likelihood of having been to MI, higher use of the car ferry to access the island and then higher use of their own car for transport on the island, a higher use of holiday houses for accommodation, and lower use of tourism information sources. Aside from these there were no differences in their responses to the sustainability initiatives or in their patterns of motivation, with the exception of seeking a luxury experience which was slightly more important for regional residents than tourists. As there were no significant differences for the variables of primary interest, these two groups were combined for all the analyses reported here.

Table 2 provides a summary of the total sample responses to the questions asking about the sustainability initiatives identified by the stakeholder group. Overall the majority of respondents were supportive of the sustainability initiatives, especially the development of more responsible tourism options, a shift to electric-powered public transport and the introduction of a National Park Fee. The lower level of support for the Solar City Initiative possibly reflected low levels of awareness, with 58% reporting that they were not aware of the program. It is worth noting that the three most popular options were those that most closely related to managing tourism's environmental impacts.

Table 3 provides a summary of responses to the four tourism scenarios. The two most appealing products were the Guided Indigenous Cultural Tours and the Self-guided Electric Bike Tours, although a substantial percentage of respondents found the other two products also appealing. In all cases at least one-fifth of the sample reported that they were willing and/or able to pay the suggested amount for the product. Taken together with the results presented in Table 2, it can be argued that there is sufficient support from both current tourists and regional residents for the MI sustainability proposals to be developed further.

Table 2. Summary of visitor support for magnetic island (MI) sustainability initiatives.

Sustainability Initiative	Negative Response	Neutral	Positive Response
The Solar City initiative was/would be important in my decision to visit MI	38%	31%	32%
The tourism operator sustainability initiatives make MI a more interesting place to visit	21%	36%	44%
I am interested to learn more about these tourism operator sustainability initiatives	28%	35%	37%
It is important that these tourism operator sustainability initiatives are further developed	10%	24%	66%
MI would be more appealing if all buses and rental cars were powered by electricity	17%	30%	53%
I would be willing to pay a $5 National Park entry fee to visit MI if the income was invested back into improving park infrastructure and facilities	25%	22%	53%
MI would be a more appealing place to visit if it was fully carbon neutral	22%	35%	44%

Figures are the percentage of respondents giving a 1 or 2 rating (Negative), a 3 rating (Neutral) or a 4–5 rating (positive), all scales were 5 points from 1 not at all to 5 very.

Table 3. Summary of visitor responses to proposed responsible tourism products.

Product Description	Overall Appeal	Make MI more attractive to visit	Encourage a longer stay	Willingness/ ability to Pay
Environmentally Friendly Camping A solar powered campground set in natural surroundings providing either a deluxe ensuite eco-tent accommodation at approximately \$130–\$170 per night or self-catering tent sites for \$20–\$30 per night with composting toilets, BBQ facilities; and hot shower facilities	29%	21%	17%	Deluxe 24% Self-catering 22%
Guided Indigenous Cultural Tours A guided 1.5 hour cultural tour with a local indigenous guide which costs \$25 for adults, \$15 for children and provides the opportunity to learn about the indigenous history and culture of the island, the natural environment of the island from an indigenous perspective, and walk to sites of local aboriginal significance	48%	42%	32%	24%
Self-Guided Electric Bike Tours A self-guided electric bike tour which costs approximately \$220 per person and includes a return ferry transfer, 1 day hire of a state-of-the-art electric bike that can travel at speeds of 25 to 30 km/h on flat ground without pedaling, a gourmet lunch, 1 night accommodation in a 3 Star resort and an Island map and information	48%	39%	34%	29%
Art/Cultural Precinct A cultural and sustainable living precinct on Magnetic Island which provides the opportunity to have a guided art safari of artist residences and galleries on the island \$30 for 2 hour tour or to participate in an art/craft workshop approximately \$45 for 2 hours and to purchase local arts, crafts and sustainable products in shops	28%	39%	29%	Safari-21% Workshop-34%

Figures are the percentage of respondents giving a 4–5 rating, all scales were 5 points from 1 not at all to 5 very.

The second step in the survey data analysis was to identify those visitors most interested in the sustainability initiatives and proposed tourism scenarios. Four sustainability market segments were identified. Committed Sustainable Visitors made up 17% of the sample and were those that scored a 4 or 5 on all the items related to the overall appeal of the four ST products and support for the general sustainability initiatives. Passive Sustainability Supporters made up 35% of the sample and they scored moderately highly on all the sustainability items. The third segment (15%) was labeled Neutral/Indifferent as they mostly chose the neutral option. The remaining 33% of the sample were those respondents who gave mostly negative ratings for the sustainability options. A closer examination of the Committed and Passive sustainability markets indicated that they were very similar in their profiles and together were significantly different to the other two groups on a number of factors. The only significant differences between these two sustainability markets were that respondents in the Committed Sustainability group were significantly more likely to say they were willing/able to pay for the ST products, to be staying longer in the region if they were tourists, to rank the opportunity to spend time in natural environments as more important to their visit to MI, and, if they did not visit MI, to report that they did not perceive it as a place that could provide the type of holiday and activities they were interested in. Visitors in Committed and Passive Sustainability segments were more likely than the other two groups to:

- Be female;
- Be first time visitors to the region if they were tourists;
- Rank learning about local culture and history and Indigenous culture, seeing wildlife and selecting environmentally and socially responsible tours and accommodation as more important to their visit MI;
- Use buses for transport on MI;
- Report being busy as a major reason for not visiting MI; and
- To be domestic tourists from the adjacent regions.

Further, the majority of respondents in both these groups (more than 70% in each group) were either regional residents or tourists who came from surrounding areas.

Overall there was support for the four tourism scenarios and the general MI sustainability initiatives suggesting that the actions could be commercially viable. In addition the strongest support came from visitors who do not have to travel far, who are interested in learning, and who are likely to stay longer both on MI and in the region. All of these are preliminary indicators of sustainability beyond MI. The results also suggested that there were low levels of awareness of the MI sustainability actions and therefore there is potential to increase visitation from markets with a stronger interest in responsible travel and recreation.

3.5. Outcomes and Future Directions

The results of the various activities within the funded project provided evidence that tourism had potential value in supporting MI residents' sustainability and QoL aspirations. An action plan was drawn up to incorporate MI stakeholder responses to the action research results as a formal finish to the funded project. The main elements of the action plan are summarized in Table 4. Subsequent projects, funded by various agencies, include the design and installation of interpretive signage, the completion of an underwater snorkel trail, refurbishment of various walking trails and the formal adoption of an educational tourism program including the development of a social marketing professional short course that uses several MI tourism opportunities and locations to highlight aspects of social marketing. Preliminary steps have also been taken towards incorporating various environmental causes, particularly related to marine turtle conservation, into events and tourism experiences. In addition to pursuing the other actions in the plan, there is also a need to further examine the potential tourist markets, evaluate projects that have been completed, and to develop and apply a more detailed pre-implementation sustainability assessment to the tourism scenarios.

Table 4. Key elements of MI tourism action plan.

Theme	Proposed Actions
Nelly Bay Terminal	Improve appearance, better use of the space to educate/inform visitors, have greater resident engagement in consultation for further planning
Enhancing Visitor Experience	Communicate the Solar City story, develop interpretive programs for sustainable lifestyles, wildlife, and environment, and interpretive signage to walks.
National Park User Fee	Investigate implementation options for user fee, identify sustainable camping options
Sustainable Transport Systems	Increase subsidies for residents, extend public transport hours, investigate costs of electric buses, improve roads, extend foot paths and bicycle paths
Revitalisation of Picnic Bay	Develop as art and culture precinct, research alternative uses of original jetty, identify and develop events
Indigenous Tours	Identify guides and potential content, increase capacity of Indigenous guides
Protect Key Assets	Reduce litter and waste, monitor effect of dredging associated with Townsville port expansion
Art Tours	Build network of artists, establish a workshop program, find a venue for workshops
Event Strategy	Develop a calendar of events, research new event ideas
Destination Marketing	Use Solar City as a marketing tool, link to slow travel, encourage longer stays, change use of images to reflect aspirations, develop educational tourism
Further Understand Visitors	Regular collection of visitors information, further investigate new/potential markets

3.6. Cautionary Notes

It is important to recognize some of the problems faced during the project. Firstly, the funded project is not sufficient on its own to move towards a new model of sustainability in tourism practice and further work is clearly necessary, although it is not clear where the support for such ongoing work will come from. The project did bring together a number of individuals and organizations and achieved some consensus, but there remains considerable mistrust and unwillingness amongst some MI residents and businesses to engage in the kinds of activities that are needed for greater community or public involvement in tourism governance. The project, especially the tourist and visitor survey, was opposed by some residents as a waste of resources, and there certainly continues to be a group of people reluctant to even discuss tourism. Some of the tourism businesses also felt that any discussion of tourism and sustainability, especially about new tourism options, implied that current activities were not sustainable and they felt the need to defend their existing track record. This particular case is also hampered by the problems of how to best deal with the physical legacy of the Nelly Bay Harbour development, which was not directly considered in this project.

There is also reluctance on the part of government and quasi-government agencies to relinquish control over tourism planning and the formal government funded tourism planning process is still very much a top down externally driven exercise. The lack of community consensus exacerbates this lack of trust in communities from government agencies and this is not an uncommon situation in community development [70]. Changes in state and national governments have also resulted in considerably less government support for community sustainability actions and this wider political context can significantly change the parameters for tourism. Taken together these factors can make the role of the academic facilitators difficult and the prospect of continued engagement in the process is a daunting one.

4. IMPLICATIONS AND CONCLUSIONS

Dunphy and Benveniste have identified six different ways in which businesses can approach sustainability [71]. The first two are rejection, where the business denies both the issues driving, and the importance of, sustainability, and non-responsiveness, referring to a lack of awareness of sustainability. The second two are compliance and eco-efficiency. Compliant businesses do what is required by regulation and/or pressure from key stakeholders and no more, while eco-efficient businesses go beyond compliance to adopt sustainability actions that can be directly linked to cost savings and efficient production. The fifth approach, called Strategic sustainability, is a logical extension of eco-efficiency which recognizes that sustainability can be used as a competitive strategy and linked to indirect business benefits. In this fifth approach the primary goal is the continued viability and profitability of the business. The final approach is Ideological Commitment or the Sustaining corporation and it is fundamentally different to the previous approaches in that the primary goal is a clear and significant contribution to the well-being of the planet and the quality of life of people in general [72]. Businesses in this approach examine not just the immediate impacts of their own business activities, but address impacts throughout the entire lifecycle of their product or service, consider the value of the product or service to those who consume it and actively work at improving sustainability beyond their business or sector [72].

Current conceptualizations of sustainable tourism can be seen as falling between the eco-efficiency and strategic sustainability approaches. This paper argues that tourism overall needs to move into the Ideological commitment stage and tourism planners, managers and businesses need to strive to become sustaining rather than just sustainable. This paper has outlined a preliminary framework for analyzing tourism that could support such a move. While this is a preliminary framework and more development is needed, the case study demonstrated that it can be used to change the way governments, communities and businesses think about tourism and sustainability.

ACKNOWLEDGMENTS

The authors wish to acknowledge the contributions and support of Sealink Qld, Townsville Enterprise, the Townsville City Council, TOBMI, MICDA, and the businesses and residents of Magnetic Island in the project described in this paper.

AUTHOR CONTRIBUTIONS

Each author contributing equally to the development, implementation and analysis of this data presented in this paper and to the writing of the paper itself.

CONFLICTS OF INTEREST

The authors declare no conflict of interest.

REFERENCES

1. Hawaiian town of Kailua asks state to stop recommending it as a tourist destination. Available online: http://mobile.news.com.au/travel/travel-updates/hawaiian-town-of-kailua-asks-state-to-stop-recommending-it-as-a-tourist-destination/story-e6frfq80-1226760222700(accessed on 13 November 2013).
2. Cohen, E. The impact of tourism on the physical environment. *Ann. Tour. Res.* **1978**, *5*, 215–237.
3. Jafari, J. Research and scholarship: The basis of tourism education. *J. Tour. Stud.* **1990**, *1*, 33–41.
4. Macbeth, J. Towards an ethics platform for tourism. *Ann. Tour. Res.* **2005**, *32*, 962–984.
5. Hardy, A.; Beeton, R.S.; Pearson, L. Sustainable tourism: An overview of the concept and its position in relation to conceptualisations of tourism. *J. Sustain. Tour.* **2002**, *10*, 475–495.
6. Saarinen, J. Traditions of sustainability in tourism studies. *Ann. Tour. Res.* **2006**, *33*, 1121–1140.
7. Jamal, T.; Camargo, B.A.; Wilson, E. Critical omissions and new directions for sustainable tourism: A situated macro-micro approach. *Sustainability* **2013**, *5*, 4594–4613.
8. Moscardo, G. Sustainable tourism innovation: Challenging basic assumptions. *Tour. Hospit. Res.* **2008**, *8*, 4–13.
9. Moscardo, G. Tourism and quality of life: Towards a more critical approach. *Tour. Hospit. Res.* **2009**, *9*, 159–170.
10. McCool, S.F. Mountains and tourism: Meeting the challenges of sustainability in a messy world. In Celebrating Mountains, Proceedings of an International Year of Mountain Conference, Jindabyne, Australia, 27 November 2002; pp. 311–318.
11. Moscardo, G. Exploring social representations of tourism planning: Issues for governance. *J. Sustain. Tour.* **2011**, *19*, 423–436.

12. Getz, D. Models in tourism planning: Towards integration of theory and practice. *Tour. Manag.* **1986**, *7*, 21–32.

13. Butler, R.W. Problems and issues of integrating tourism development. In *Contemporary Issues in Tourism Development*; Pearce, D.G., Butler, R.W., Eds.; Routledge: London, UK, 1999; pp. 65–80.

14. Wall, G. Is ecotourism sustainable? *Environ. Manag.* **1997**, *21*, 483–491.

15. Moscardo, G. Challenges for tourism development on Australia's Great Barrier Reef coast. In*Coastal Tourism Development—Planning and Management Issues*; Dowling, R., Pforr, C., Eds.; Cognizant Communications: New York, NY, USA, 2009; pp. 252–264.

16. Jafari, J. Bridging out, nesting afield: Powering a new platform. *J. Tour. Stud.* **2005**, *16*, 1–5.

17. Jovicic, D.Z. Key issues in the implementation of sustainable tourism. *Curr. Issues Tour.* **2013**.

18. Holden, A. The environment-tourism nexus—Influence of market ethics. *Ann. Tour. Res.***2009**, *36*, 373–389.

19. Leslie, D. Introduction. In *Responsible Tourism: Concepts, Theory and Practice*; Leslie, D., Ed.; CABI: Wallingford, UK, 2012; pp. 1–16.

20. Glossary of statistical terms. Available online: http://stats.oecd.org/glossary/detail.asp?ID=2218 (acces sed on 1 December 2013).

21. Moscardo, G. Building social capital to enhance the Quality-of-life of destination residents. In *Handbook of Tourism and Quality-of-Life Research*; Uysal, M., Perdue, R., Sirgy, M.J., Eds.; Springer: New York, NY, USA, 2012; pp. 403–422.

22. Costanza, R.; Fisher, B.; Ali, S.; Beer, C.; Bond, L.; Boumans, R.; Danigelis, N.L.; Dickinson, J.; Elliott, C.; Farley, J.; *et al.* Quality of life: An approach integrating opportunities, human needs, and subjective well-being. *Ecol. Econ.* **2007**, *61*, 267–276.

23. Malkina-Pykh, I.G.; Pykh, Y.A. Quality-of-life indicators at different scales: Theoretical background. *Ecol. Indicat.* **2008**, *8*, 854–862.

24. Lehtonen, M. The environmental-social interface of sustainable development: Capabilities, social capital, institutions. *Ecol. Econ.* **2004**, *49*, 199–214.

25. Vermuri, A.W.; Costanza, R. The role of human, social, built and natural capital in explaining life satisfaction at the country level: Toward a national well-being index. *Ecol. Econ.* **2006**, *58*, 119–133.

26. Moscardo, G.; Lamberton, G.; Wells, G.; Fallon, W.; Lawn, P.; Rowe, A.; Humphrey, J.; Wiesner, R.; Pettitt, B.; Clifton, D.; *et al. Sustainability in Australian Business: Principles and Practice*; Wiley: Brisbane, Australia, 2013.

27. Moscardo, G.; Konovalov, E.; Murphy, L.; McGehee, N. Mobilities, community well-being and sustainable tourism. *J. Sustain. Tour.* **2013**, *21*, 532–556.

28. Macbeth, J.; Carson, K.; Northcote, J. Social capital, tourism and regional development: SPCC as a basis for innovation and sustainability. *Curr. Issues Tour.* **2004**, *7*, 502–522.

29. McGehee, N.; Lee, S.; O'Bannon, T.; Perdue, R. Tourism-related social capital and its relationships with other forms of capital: An Exploratory study. *J. Trav. Res.* **2010**, *49*, 486–500.

30. Andereck, K.; Nyuapane, G. Exploring the nature of tourism and quality of life perceptions amongst residents. *J. Trav. Res.* **2010**, *50*, 248–260.

31. Scott, D. Why sustainable tourism must address climate change. *J. Sustain. Tour.* **2011**, *19*, 17–34.

32. Lansing, P.; de Vries, P. Sustainable tourism: Ethical alternative or marketing ploy? *J. Bus. Ethics* **2007**, *72*, 77–85.

33. Hall, C.M. Policy learning and policy failure in sustainable tourism governance: From first- and second-order to third-order change? *J. Sustain. Tour.* **2011**, *19*, 649–671.

34. Ruhanen, L. Strategic planning for local tourism destinations: An analysis of tourism plans. *Tour.Hospit. Plann. Dev.* **2004**, *1*, 239–253.

35. Ko, T.G. Development of a tourism sustainability assessment procedure: A conceptual approach. *Tour. Manag.* **2005**, *26*, 431–445.

36. Garcia-Rosell, J.; Makinen, J. An integrative framework for sustainability evaluation in tourism: Applying the framework to

tourism product development in Finnish Lapland. *J. Sustain. Tour.* **2013**, *21*, 396–416.

37. Tao, T.C.H.; Wall, G. Tourism as a sustainable livelihood strategy. *Tour. Manag.* **2009**, *30*, 90–98.

38. Truong, V.D.; Hall, C.M. Social marketing and tourism: What is the evidence? *Soc. Market. Q.* **2013**, *19*, 110–135. [Google Scholar]

39. Marzuki, A.; Hay, I. Towards a public participation framework in tourism planning. *Tour. Plann. Dev.* **2013**.

40. Moscardo, G. Community capacity building—An emerging challenge for tourism development. In *Building Community Capacity for Tourism Development*; Moscardo, G., Ed.; CABI: Wallingford, UK, 2008; pp. 1–15.

41. Botes, L.; van Rensburg, D. Community participation in development: Nine plagues and twelve commandments. *Commun. Dev. J.* **2000**, *35*, 41–68.

42. Feighery, W.G. Consulting ethics. *Ann. Tour. Res.* **2011**, *38*, 1031–1050.

43. Moriarty, J.P. Theorising scenario analysis to improve future perspective planning in tourism. *J. Sustain. Tour.* **2012**, *20*, 779–800.

44. Coria, J.; Calfucura, E. Ecotourism and the development of indigenous communities: The good, the bad, and the ugly. *Ecol. Econ.* **2012**, *73*, 47–56.

45. Department of National Parks, Recreation, Sport and Racing. About Magnctic Island. Available online: http://www.nprsr.qld.gov.au/parks/magnetic-island/about.html (accessed on 11 November 2013).

46. Australian Bureau of Statistics. National regional profile: Magnetic Island. Available online:http://www.abs.gov.au/ausstats/abs@nrp.nsf/781eb7868cee03e9ca2571800082bece/2a3e7368b29b7557ca2577d50011bdaa!OpenDocument (accessed on 11 November 2013).

47. Pure Magnetic. Island History. Available online: http://www.puremagnetic.com/page.asp?CID=magnetic-island-history (accessed on 11 November 2013).

48. Australian Heritage. Magnetic Island. Available online: http://www.heritageaustralia.com.au/search.php?state=QLD&region=93&view=1232 (accessed on 11 November 2013).

49. Harrington, J.T. 'Being here': Heritage, Belonging and Place Making. Ph.D. Thesis, James Cook University, Townsville, Australia, 2004.

50. Dickie, P. Nelly Bay. Available online: http://www.rag.org.au/phildickiestories/Walkden.htm(accessed on 12 November 2013).

51. Heywood, P. Social justice and planning for the public interest. *Urban Pol. Res.* **1990**, *8*, 60–68.

52. Great Barrier Reef Marine Park Authority. Nelly Bay Harbour Development. Available online: http://www.gbrmpa.gov.au/outlook-for-the-reef/great-barrier-reef-outlook-report/outlook-online?sq_content_src=%2BdXJsPWh0dHAlM0ElMkYlMkZ3d3ctcmMuZ2JybXBhLmdvdi5hdSUyRmNvcnBfc2l0ZSUyRm1hbmFnZW1lbnQlMkZlYW0lMkZlaWElMkZuZWxseV9iYXkmYWxsPTE%3D (accessed on 13 November 2013).

53. Valentine, P. Community response to tourism proposals on Magnetic Island. *Urban Pol. Res.* **1989**, *7*, 83–85.

54. Moyle, B.B.; Croy, G.; Weiler, B. Community perceptions of tourism: Bruny and Magnetic Islands, Australia. *Asia Pac. J. Tour. Res.* **2010**, *15*, 353–366.

55. Asher, C.; Roe, D.; Goodwin, H. *Pro-Poor Tourism Strategies: Making Tourism Work for the Poor*; Overseas Development Institute: Nottingham, UK, 2001.

56. Blackstock, K. A critical look at community based tourism. *Commun. Dev. J.* **2005**, *40*, 39–49.

57. Dredge, D. Place change and tourism development conflict: Evaluating public interest. *Tour. Manag.* **2010**, *31*, 104–112.

58. Holladay, P.J.; Powell, R.B. Resident perceptions of social-ecological resilience and the sustainability of community-based tourism development in the Commonwealth of Dominica. *J. Sustain.Tour.* **2013**.

59. Lee, T.J.; Riley, M.; Hampton, M.P. Conflict and progress: Tourism development in Korea.*Ann. Tour. Res.* **2010**, *37*, 355–376.

60. Park, D.B.; Lee, K.W.; Choi, H.S.; Yoon, Y. Factors influencing social capital in rural tourism communities in South Korea. *Tour. Manag.* **2012**, *33*, 1511–1520.

61. Townsville City Council. Townsville tourism industry. Available online: http://www.townsville.qld.gov.au/townsville/tourism/Pages/industry.aspx (accessed on 16 November 2013).

62. Tourism Queensland. *Destination Management Plan for Tourism in Townsville North Queensland 2007–2010*; Tourism Queensland: Brisbane, Australia, 2007.

63. Tourism Queensland. *Townsville North Queensland Tourism Opportunity Plan 2009–2019*; Tourism Queensland: Brisbane, Australia, 2009.

64. Townsville Solar City. Magnetic Island Solar Suburb. Available online: http://www.townsvillesolarcity.com.au/Project (accessed on 16 November 2013).

65. Bruce, A.; Heslop, S.; Macgill, I.; Watt, M. *Magnetic Island and Townsville Solar City*; Australian Renewable Energy Agency: Canberra, Australia, 2013.

66. Isaac, J. Smart living on Magnetic Island. *Ecos* **2009**, *148*, 9.

67. Ergon Energy. Magnetic Island's successful Solar City project wraps up. Available online: http://www.ergon.com.au/about-us/news-room/media-releases/regions/northern/magnetic-islands-successful-solar-city-project-wraps-up (accessed on 16 November 2013).

68. Citiwater Townsville. *Magnetic Island Water Recycling*; Citiwater: Townsville, Australia, 2002.

69. Tourism Queensland. *Sustainable Regions for Queensland—Magnetic Island*; Tourism Queensland: Brisbane, Australia, 2009.

70. Eversole, R. Managing the pitfalls of participatory development: Some insight from Australia. *World Dev.* **2003**, *31*, 781–795.

71. Dunphy, D.; Benveniste, J. An introduction to the sustainable corporation. In *Sustainability: The Corporate Challenge of the 21st Century*; Dunphy, D., Benveniste, J., Griffiths, A., Sutton, P., Eds.; Allen & Unwin: Sydney, Australia, 2000; pp. 3–18.

72. Benn, S.; Dunphy, D.; Griffiths, A. Enabling change for corporate sustainability: An integrated perspective. *Australas. J. Environ. Manag.* **2006**, *13*, 156–165.

8

CHAPTER

Towards Sustainable Tourism Development in Urban Areas: Case Study on Bucharest as Tourist Destination

Andreea Zamfir [†,*] and Razvan-Andrei Corbos [†]

Faculty of Management, The Bucharest University of Economic Studies, Piata Romana 6, Bucharest 030173, Romania

ABSTRACT

The issues of tourism sustainability and urban development have become major priorities for public policy makers across the globe. Today, maybe more than ever, there is a need for managing sustainable tourism development, and this cannot be attained without taking into account environmental problems and their global dimension. Various problems and requirements of society and of the development of urban areas may be solved by transforming the cities into attractive tourist destinations. Therefore, this study explores how sustainable tourism development in urban areas can be basically achieved and managed. The paper discloses some success factors for managing sustainable tourism development in urban areas and emphasizes a case study regarding Bucharest, the capital of Romania, as a tourist destination. The originality and value of this study consist of identifying the main ways of developing sustainable tourism in Bucharest based upon empirical research conducted with the aid of a survey. The findings of this study may be helpful for upcoming research in the area of managing sustainable urban tourism development.

Keywords: sustainable tourism development; urban areas; urban development; tourist destination; tourism management

1. INTRODUCTION

Nowadays, maybe more than ever, there is a need for managing sustainable tourism development, and this cannot be attained without taking into account environmental problems and their global dimension. There are many problems and needs of society and of the development of urban areas that may be solved by transforming the cities into attractive tourist destinations. Therefore, the aim of this study is to explore how sustainable tourism development in urban areas can practically be achieved and managed.

Sustainable development is a multidimensional concept, which is related to the environment and resources as well as to the population and industrial and agricultural production [1]. A sustainable behavior of consumption is the key element of long-term development [2] (p. 143). Therefore, achieving sustainable development implies solutions aiming at the same time at the social, political, economic, demographic, and technical fields [1]. Sustainable development may be achieved if it relies on the harmony between economic progress and the boundaries of nature, particularly between the quantity and the regeneration time of natural sources, human-made emissions, and the neutralization capabilities of nature [3].

Therefore, to be more specific, the focus and purpose of this study is to identify the main success factors for managing sustainable tourism development in urban areas, based on a literature review and on a case study regarding Bucharest as a tourist destination. The research was conducted using evidence provided by conceptual papers, research papers, and literature reviews, as well as books and internet sources. In order to answer the research question, we have first revealed some modern approaches in defining sustainable tourism. Secondly, the study disclosed some success factors for managing sustainable tourism development in urban areas. Thirdly, we have emphasized a case study regarding Bucharest, the capital of Romania, as a tourist destination with an eye to identify the main ways of developing sustainable tourism in Bucharest based upon empirical research conducted with the aid of a survey. In order to answer the research question, we have analyzed the published sources, and we have evaluated and interpreted the empirical data collected through the questionnaire created.

The contribution this paper will make to the literature and practice derives from the logical connections of the arguments and counter-arguments for identifying the success factors for managing sustainable tourism development in urban areas, as well as from the possibility of using the findings in comparative urban studies or in order to conceive a strategy for developing Bucharest as a sustainable tourist destination.

2. MODERN APPROACHES IN DEFINING SUSTAINABLE TOURISM

There are many approaches and ways of defining the concept of sustainable tourism that can be found in the literature, and many of them start the inquiry by making a transition from the concepts of sustainability or sustainable development. The abundance of definitions, meanings, and implications will be further reviewed in order to reveal the essence of the sustainable tourism concept.

Sustainable development may be defined as an all-embracing concept which involves all aspects of human activity and in which all the nations of the world should be involved in Zaharia *et al.* [4] (p. 159). Moreover, it is seen as one of the major challenges of modern evolution. In order to achieve sustainable development, there is a need to identify the main causes of environmental degradation firstly, to assess its size secondly, to find reasonable solutions to counteract it thirdly, and to reduce its negative impacts on economic and social life fourthly [5] (p. 99).

As Rifkin [6] asserts, a significant impact into the 21st century will be the Third Industrial Revolution, which will radically change all facets of working and living. The social, economic, and political life of the previous industrial revolutions based on fossil fuels was characterized by a conventional top-down organization of society. However, this type of organization is rapidly changing, and new distributed and collaborative relations emerge as a foundation of the green industrial era [6].

The concept of sustainable tourism emerged in the early 1990s, by connecting sustainable development ideas and principles with tourism. This new concept, at that time, was quickly accepted and promoted by many international and national organizations. For instance, the International Union for Nature Conservation, World Federation for Nature Protection, and the European Federation of National and Natural Parks defined sustainable tourism in 1991 as development, marketing, and management of all forms/types of tourism taking into account the natural, social, and economic environment and preserving the natural and cultural resources for future generations [7,8]. From this definition it follows that tourism (ecotourism, green tourism, rural tourism, business tourism, urban tourism, *etc.*) should be based on the principles of sustainable development [9] (p. 250). Currently, there is a strengthening of political concerns for nature conservation due to the aggravation of environmental problems and due to the acknowledgment of the importance of maintaining the quality of the environment for future generations [5] (p. 100).

According to the World Tourism Organization, the development of sustainable tourism responds to the present needs of tourists and to those of the host regions, and also protects and increases future chances and opportunities. Through sustainable tourism, all resources can be managed so as to satisfy needs and to maintain cultural integrity, biological diversity, ecological dimensions, and the life system [10]. Accordingly, sustainable tourism can be defined as a form of tourism that takes into account present and future impacts (economic impact, social impact, and environmental impact), and responds to the various needs of tourists, industry, and local communities, as well as the environment [11] (p. 85).

Tourism sustainability is a complex concept because it has a latent, relative, and multidimensional nature [12] (p. 277). The multidimensional nature of tourism sustainability is based on three distinct sustainability dimensions: environmental, social, and economic sustainability [13] (pp. 5–6). As a result, sustainable tourism must undertake the environmental, social, and economic principles of sustainability (Figure 1) [14] (p. 3). This way the synergy effect is produced at the level of tourism sustainability. As a general rule, synergy means that certain features occur within the whole system without occurring in any of the subsystems [15] (p. 68).

Sustainable tourism covers all types/forms of tourism: conventional mass tourism, cultural tourism, mountain tourism, seaside tourism, spa tourism, business tourism, medical tourism, rural tourism, urban tourism, *etc.* Regardless of the type or tourism activities, they should take on the principles of sustainable tourism: (1) the local community should initiate the tourism activity by its own means, maintaining control over tourism development in the area; (2) tourism should provide jobs to the residents, thus improving the quality of life; (3) the accepted international standards in tourism should be used when the guidelines for tour operators, for monitoring the impact of various tourism activities, and for setting the acceptability limits for different areas are established; (4) educational and training programs to improve management in the area of protecting natural and cultural resources must be implemented [16].

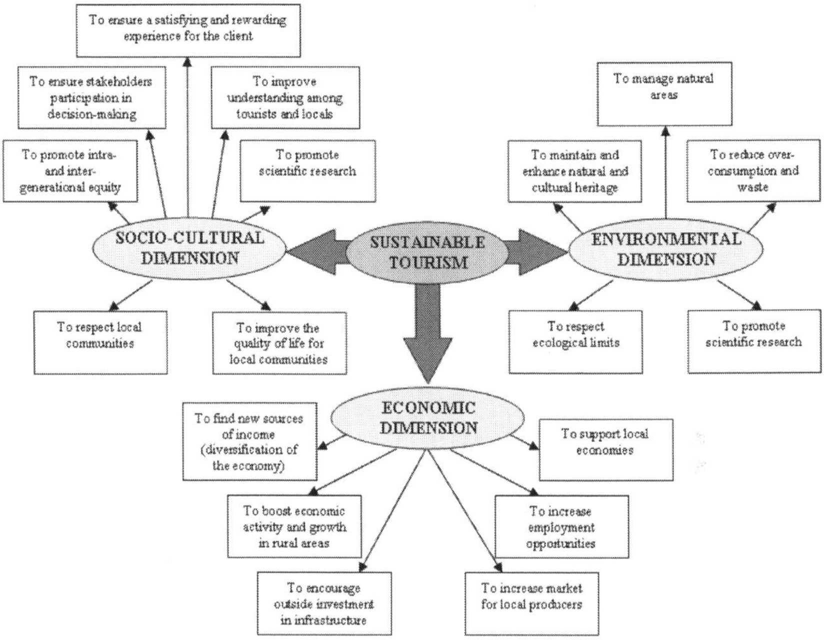

Figure 1. Principles of sustainable tourism. Source: Adapted from White *et al.* [14] (p. 4).

3. SUCCESS FACTORS FOR MANAGING SUSTAINABLE TOURISM DEVELOPMENT IN URBAN AREAS

Nowadays, cities must assume an important role in addressing environmental problems. They are innovation, efficiency, investment, and productivity centers, but they are also an important cause of climate change. The solutions to sustainable development and global climate change are to be found in the ecological urbanization of cities. Therefore, the campaigns and projects carried out by urban managers in order to reach the sustainable development of regions, cities, and rural areas around are becoming extremely important these days [9] (p. 254).

The role of urban areas in sustainable development is more and more recognized at a global level. Providing services to residents and businesses, creating jobs, stimulating research, and development within and outside the economic sector are only some of the functions of the cities. However, these functions are not only limited to administrative boundaries of cities, but they also create benefits for the whole region in which cities are included. A success condition for a modern city is to demonstrate that it meets the environmental requirements [17] (p. 379).

Even if local socio-economic development is achievable and accelerated in urban areas, these are, at the same time, places of environmental problems [18] (p. 79). Cities are special local environments, and the manner in which they should be governed and managed is determined by their features [19] (p. 31). A feasible urban development strategy should include elements relating to the image and brand of the city, elements which will ensure a competitive advantage for the local community [20] (p. 734). Today, business organizations and not-for-profit organizations, among which is the government, need to create and maintain permanent and effective mechanisms which to enable them to adjust in a timely manner in order to respond to trend changes that characterize the environment in which they perform [21] (p. 7).

In this context, the local authorities, along with other partners (stakeholders), may establish the strategy for sustainable urban development and the suitable forms of assistance [17] (p. 383). Today, a new paradigm is emerging, in which the individual is recognized as a partner of the public administration, being referred to as a "stakeholder" [22] (p. 41). This partnership of urban stakeholders may lead to the development of urban areas through integrated urban transport systems for the accessibility and mobility of citizens, as well as environmentally friendly buildings that use renewable energy [17] (p. 383). Nowadays cities are operating on a global market, competing with other urban areas around the world for investments, residents, and tourists, and thus there is a need for competitiveness [17] (380). When tourists are choosing their holidays packages they make comparisons, evaluate the possible destinations, and estimate the benefits and drawbacks of each one as a potential holiday destination [23] (p. 808). The tourists have the liberty to set up their personal itineraries to follow; however, they are not free to interfere in the natural course of the environment they are visiting [24].

The negative ecological, economic, and social effects of tourism may be lessened through professional urban management [10] (p. 256). It is important to mention that [19] has identified a relatively clear distinction between urban management, with reference to the officials executing the policies, and urban governance, which refers to additional components [19] (p. 34). The author has concluded in his article that urban management is a reform of city administration and that its task is to create equilibrium between social and economic development, which have a fragile coexistence [19].

Tourism enhancement implies planning, development, management, and continuous improvement [25]. The continuous improvement process should be applied to all forms of tourism, whether we are referring to urban tourism, resorts, rural and coastal areas tourism, or mountains and protected areas tourism [12]. Recreational activities may be developed within natural areas and may bring significant revenues both to those who manage them and to the local communities. Tourism is one of the few economic sectors through which sustainable development of these areas can be achieved, and ecotourism is the most accepted form of sustainable tourism for any country or region of the world [10] (p. 256). Small tourism enterprises have the potential to help tourism destinations to progress towards sustainability objectives because of their numerical dominance, their central role in human activities, and their increasing importance within the framework of sustainable tourism development [26] (p. 575).

One of the main barriers to achieving sustainable tourism refers to the difficulty of measuring the sustainability level of a tourism destination. This barrier has hindered the decision-making processes and made it difficult to encounter the needs of the tourism destinations [12] (p. 278). Achieving sustainable tourism means taking the effects and needs into account in its planning and development [11] (p. 85).

Furthermore, one of the main challenges in achieving sustainable tourism is to fill the existing gap between the stages of planning and implementation. The stage of planning includes the design of methodologies, tourism policies, and technological know-how. The stage of implementation refers to the execution of tourism projects and plans by public agencies and tourism companies [12] (p. 279).

In order to plan and manage sustainable tourism development there is a need for thorough insight into the present levels of sustainability or at least sustainability performance, because this is a precondition for designing effective and efficient policies [13] (p. 4). The definitions of sustainability indicators must take into account the interconnectivity in the tourism system and must reflect the distinctive environmental, economic, and socio-cultural attributes of the destination or entity to which they would be applied [26].

Development of information and communication technologies has created the premises for better collaboration and communication between tourism organizations, on one hand, and between them and their clients (the tourists), on the other hand [27] (pp. 84–85). According to [28], it is predicted that the Internet will be the key to future management of the tourism sector, in the sense that only the companies that will use the technology to identify the customer needs and to respond to them will survive on the market [28] (p. 501). There is a need to apply some innovative technologies within the tourism sector, since certain services are still based on outdated and unsuccessful methods. The investments should focus on developing the infrastructure in the tourism sector, mainly the major structures such as airports, retail centers, theaters, museums, hotels, and conference centers, which may influence the visible and invisible impact of the tourism sector on the environment [24] (p. 127). Tourism companies with a well-designed website have access to a vast tourism market, and they can offer their services to a wide range of clients both at the national and international level [28] (p. 492).

Information and communication technologies may be used by the companies to increase service flexibility and adaptability to the needs of the clients [27] (pp. 84–85). For instance, consistent with [29] (p. 133), through the Web, the museum "breaks" the traditional value chain in the sense that it enhances its role from being just a "commodity" with value added to the local tourism services to becoming the catalyst of an innovative "niche" supply developed around the cultural event. Moreover, if effectively integrated in the service process, new technologies could actively support the development of strategies related to innovation, collaboration, and value co-creation, playing a key role in providing competitive services [27] (p. 85).

4. CASE STUDY REGARDING BUCHAREST AS A TOURIST DESTINATION

4.1. Data and Methodology

Bucharest, the capital of Romania, was included in the online contest "The Best City to Visit Travel Tournament 2013 Championship" initiated by the site Foxnomad to designate the city visited in 2013 by the blogger Anil Polat [30]. Even though Bucharest did not win the final online contest, it still remained the second destination to visit according to the voters. This is not surprising if we are taking into account the variety of places of interest for tourists in Bucharest such as museums, parks, administrative buildings, restaurants, clubs, *etc.* The Romanian capital fought hard against and defeated Aswan, Rome, Cape Town, Edinburgh, and Granada and it was finally eliminated by another Romanian city, Sibiu [30].

The urban essence that makes Bucharest a unique city, with its strengths, weaknesses, opportunities, and threats, must be found in business urban life, as well as in resident and tourist life [31] (p. 277). Therefore, the approach of the sustainable urban tourism development should start from the people, from the residents and tourists. They should be partners of the government, of the public administration, of the policy-makers. Consequently, we assert that the opinion of residents and tourists matters because they are the beneficiaries of the public and private services provided. When choosing a destination, the tourists are searching for those services that best suit their needs.

Furthermore, we argue that any policy or strategy in the area of sustainable urban tourism development should start from the needs and desires of the people and should be connected to the principles of sustainable development. Thus, we assert that the method we used in our research was the best approach in this context.

In order to identify people's opinions regarding the main ways of developing sustainable tourism in Bucharest, we have created a questionnaire which was applied between 14–17 March 2013 at Romania's Tourism Fair. The questionnaire used was constructed based on a five-level Likert scale of increasing intensity where 1 indicates strongly disagree (very badly) and 5 designates strongly agree (very good). The sampling method was random and 236 subjects, residents aged 18 and over, participated in the research.

Most of the respondents are aged between 18 and 25 years (34% of the respondents), followed by respondents aged between 36 and 45 years (26% of the respondents), respondents aged between 26 and 35 years (25% of the respondents), respondents aged between 46 and 55 years (11% of the respondents), and respondents aged over 55 years (4% of the respondents). Regarding gender, 56% of the respondents are female, while 44% of them are male. The research adequately reflects gender distribution in Bucharest in statistical terms.

The information collected was processed using Microsoft Excel 2003, both for centralizing and analyzing data.

The next section of the article briefly presents the results of the survey, emphasizing people's opinions regarding the main attractions of Bucharest as tourist destinations, recreational facilities and sources of information about them, institutions involved in promoting sustainable tourism in urban areas, and the role of education in promoting sustainable urban tourism.

4.2. Results of the Survey

4.2.1. Tourist Attractions and Places to Visit

With reference to the main tourist attractions of Bucharest (Figure 2), among the responses to the question "What comes to mind when you think about the main attractions of Bucharest as a tourist destination?" the most common answer of the respondents was "Palace of Parliament (People's House)", with 21% of all persons surveyed. In second place came "Dimitie Gusti National Village Museum", with 13% of the respondents, followed by "Grigore Antipa Natural History Museum", with 12% of the respondents. Less than 5% of persons surveyed associated the attractions of Bucharest with churches or parks.

For example, Dimitrie Gusti National Village Museum has a website only in the Romanian language. Therefore, we strongly recommend the translation of museums' official websites not only in English, but also in other foreign languages, so as to increase information availability and to attract a bigger number of foreign visitors. Furthermore, museums may rely on multimedia information systems in order to redesign their traditional services and to promote new cultural experiences. Multimedia information systems allow museums to involve a worldwide network of potential visitors in the production of the cultural service [32] (p. 407).

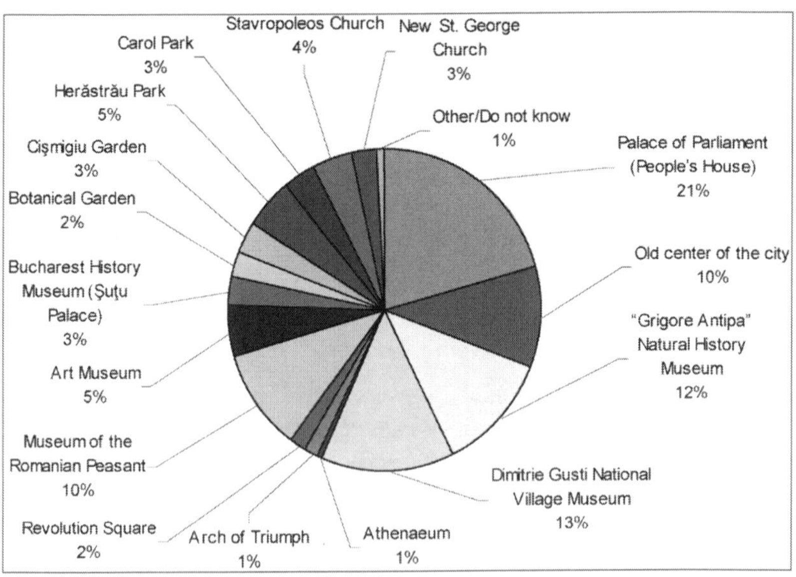

Figure 2. Main tourist attractions in Bucharest. Source: Authors, based on the survey.

Organizing different festivals and other cultural events in the city may be another way of attracting tourists [33] (p. 19).

Regarding the preferences of the respondents (Figure 3), among the responses to the question "What do you like to visit in Bucharest?" the most common answer of the respondents was "museums", with 45% of the responses, followed by "buildings" together with "historical center", each accounting for 19% of respondents. At the opposite pole, the "parks" are preferred only by 4% of the respondents.

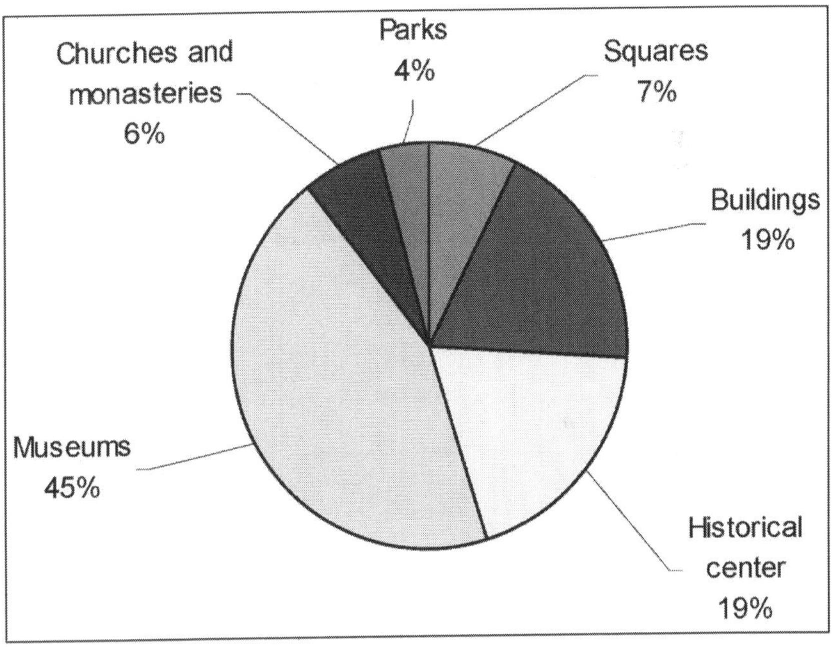

Figure 3. Types of places chosen by the respondents to visit in Bucharest. Source: Authors, based on the survey.

The heterogeneousness of the Romanian museums may help in playing an important role in educating visitors as long as there are appropriate management and marketing strategies implemented. Moreover, these strategies may benefit the development of information and communication technologies so as to lead to providing competitive cultural services [32] (p. 408).

On the topic of the recreational facilities (Figure 4), 10% of the respondents appreciate them as "very good", 43% as "good", 22% as "neither good nor bad", 19% as "bad", and 6% as "very bad".

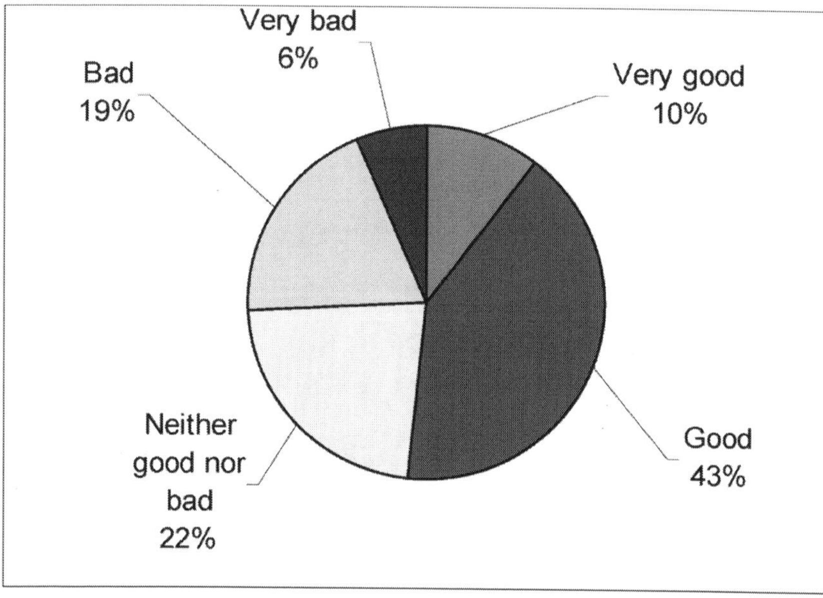

Figure 4. Recreational facilities in Bucharest. Source: Authors, based on the survey.

4.2.2. Sources of Information

The Internet is the main source of information regarding tourist destinations (Figure 5) for the respondents participating in the study, with 88% of them choosing it. Way behind as information source are television (9%), radio (2%), and newspapers (1%).

4.2.3. Ways of Developing Sustainable Tourism in Bucharest

Regarding the role of institutions in promoting sustainable tourism in urban areas (Figure 6), a majority of respondents (60%) consider that the public administration (regional and local public authorities –36% and central public authorities –24%) are in the best position to implement programs to promote and develop sustainable tourism in urban areas. In second place came the non-governmental organizations (19% of the respondents), followed by citizens (16% of respondents).

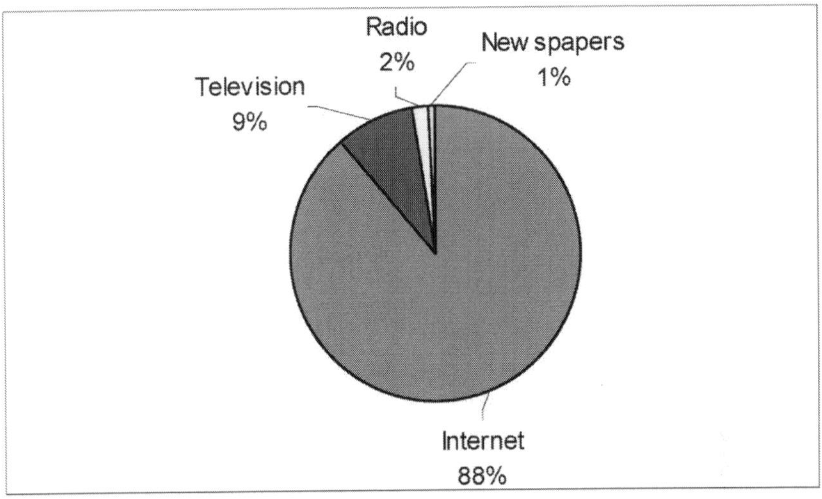

Figure 5. Sources of information regarding tourist destinations. Source: Authors, based on the survey.

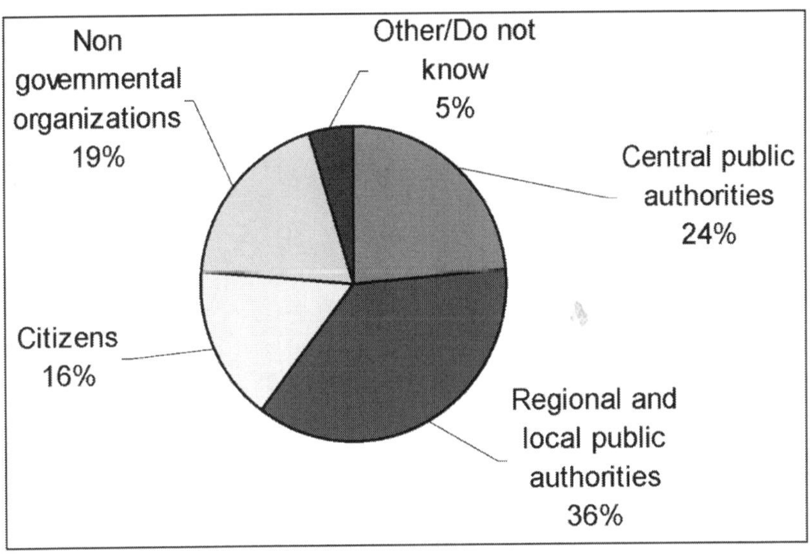

Figure 6. Institutions involved in promoting sustainable tourism in urban areas. Source: Authors, based on the survey.

Public authorities may create an urban brand in order to attract tourists, and to convince them to recommend the city to other visitors [34] (p. 132).

According to the performed study, the educational activities on environmental issues contribute to achieving sustainable urban tourism for 60% of the respondents, of which 11% strongly agree and 49% agree (Figure 7).

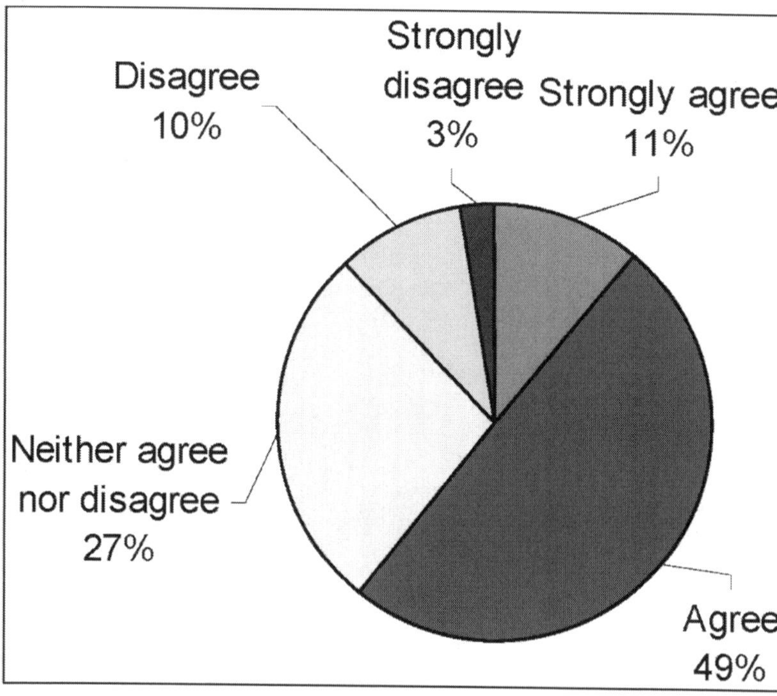

Figure 7. Important role of education in achieving sustainable urban tourism. Source: Authors, based on the survey.

The main ways of developing sustainable tourism in Bucharest (Figure 8) are the limitation of waste and environment degradation, as well as the integration of tourism into planning (both with 18% of respondents). In addition, applying better management based on sustainable urban tourism strategies is seen as a way of developing sustainable tourism in Bucharest by 15% of the respondents, followed by educating people and maintaining the natural and cultural heritage (both with 14% of the respondents).

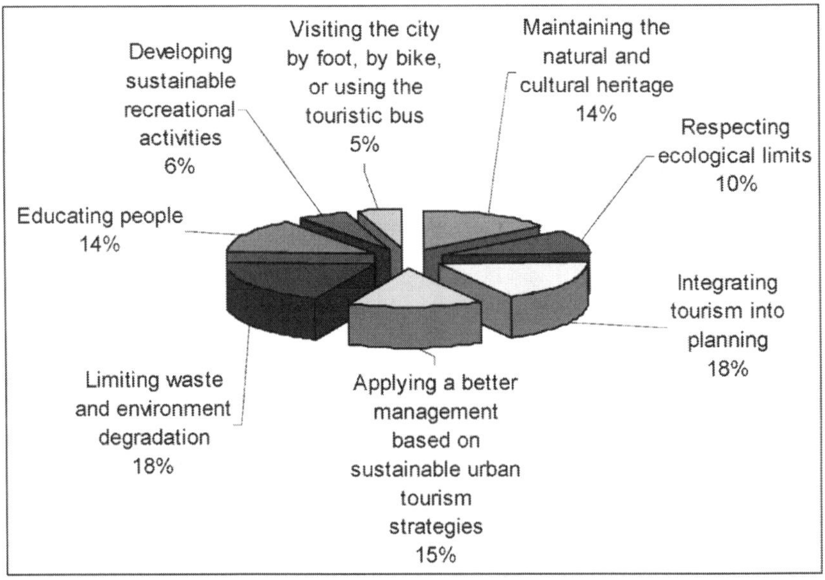

Figure 8. Ways of developing sustainable tourism in Bucharest. Source: Authors, based on the survey.

Ecological education aims to increase the awareness and understanding of the relationships between people and nature in order to maintain a balance between the present requirements and the future needs [5].

Also, sustainable tourism in Bucharest may be developed through respecting ecological limits (10% of the respondents), developing sustainable recreational activities (6% of the respondents), and visiting the city by foot, by bike, or using the touristic bus (5% of the respondents).

The public authorities may analyze the deficiencies and barriers faced in the process of strategic planning of sustainable tourism in Bucharest, and they may identify the main problems and may indicate the development priorities.

5. DISCUSSION

The results of the survey reveal that there are some points worth highlighting. On one hand, the capital city may be a good destination for cultural tourism due to the heterogeneousness of the museums and other cultural places and events (festivals, concerts, *etc.*) that can be organized. On the other hand, there are many actions that need to be done in order to promote Bucharest as a cultural tourist destination. For instance, increasing information availability, the translation of official websites into English and other foreign languages, and using multimedia information systems are only some of the recommendations to redesign the traditional services and to promote new cultural experiences for tourists.

Another idea emerging from the data analyzed is that the Internet is the main source of information regarding tourist destinations. Therefore, we strongly recommend using the advantages of the Internet to attract tourists, and to further involve them in the production of tourist services. Social networks like Twitter or Facebook may be used to promote Bucharest as a tourist destination due to rapid dissemination of information. Not unintentionally, we have presented the results of the blog online contest "The Best City to Visit Travel Tournament 2013 Championship" in the beginning of Section 4 of this article. Such online contests or ways of interaction attract people, and may be used within the strategies to promote tourist urban destinations.

The public administration is in the best position to implement programs, policies, and strategies to promote and develop sustainable tourism in urban areas. These strategies may benefit the development of information and communication technologies so as to lead to providing competitive services. Public authorities may create an urban brand in order to attract tourists and to convince them to recommend the city to other visitors.

The main ways of developing sustainable tourism in Bucharest identified by the respondents suggest that they care about the environment, as well as about the natural and cultural heritage.

On the basis of what has been revealed above, we may bring the paper to a close by saying that it may be appropriate to find a way to apply better management based on sustainable urban tourism strategies. The public administration, the residents, and the private companies, as well as other stakeholders, play an important role in this endeavor. The development of Bucharest as a sustainable tourist destination is possible only if there is a real partnership between all actors more or less involved in tourist activities.

6. CONCLUSIONS

This study has revealed some success factors for managing sustainable tourism development in urban areas and has emphasized a case study regarding Bucharest, the capital of Romania, as a tourist destination. The main conclusions of this study may be summarized as follows: (1) among the success factors for managing sustainable tourism development in urban areas, the strategic planning plays an important role; (2) sustainable tourism development in Romania may be achieved through developing urban areas as touristic destinations; and (3) the subjects comprised in this study appreciate that sustainable tourism development in Bucharest may be achieved in different manners.

The originality and value of this study consist of identifying the main ways of developing sustainable tourism in Bucharest based upon empirical research conducted with the aid of a survey. These ways of developing sustainable tourism in Bucharest are the following: limitation of waste and environment degradation, integration of tourism into planning, practice of better management based on sustainable urban tourism strategies, education, preservation of the natural and cultural heritage, avoidance of exceeding ecological limits, and development of sustainable recreational activities. Any policy or strategy in the area of sustainable urban tourism development should start from the needs and desires of the people and should be connected to the principles of sustainable development.

Our current experience advises us that there is no panacea for successful sustainable urban tourism development. The findings of this study can inform future research on urban sustainability by using them in comparative studies or as a starting point for developing policies and strategies in the field. We suggest that further research should be done in the direction of identifying the opinions of the representatives of the public administration and of tourism organizations, and to use them along with the findings of this study to conceive a strategy for developing Bucharest as sustainable tourist destination. Another direction for future research could be to replicate the survey in various cities from different regions in order to find out how local culture may influence the approach of sustainable urban tourism and what its effects are (if any) on the development preferences.

AUTHOR CONTRIBUTIONS

Andreea Zamfir and Razvan-Andrei Corbos designed the research and identified the logical connections of the arguments and counter-arguments; Andreea Zamfir collected the data; Razvan-Andrei Corbos organized the data; Andreea Zamfir and Razvan-Andrei Corbos analyzed the data; Andreea Zamfir and Razvan-Andrei Corbos wrote the paper. All authors read and approved the final manuscript.

CONFLICTS OF INTEREST

The authors declare no conflict of interest.

REFERENCES

1. Lazar, C.; Lazar, M. The Quantification of the Sustainable Development at Local Level. *WSEAS Trans. Bus. Econ.* **2008**, *5*, 310–319.

2. Marinescu, P.; Burcea, M. Information and Ecological Behaviour towards the Natural Resources Consumption of the Population of Bucharest. *Amfiteatru Econ.* **2012**, *XIV*, 142–156.

3. Pozeb, V.; Krope, T. Importance of Legal Protection and International Quality Standards for Environmental Protection. In IASME/WSEAS, Proceedings of the 2nd IASME/WSEAS International Conference on Energy & Environment, Portoroz, Slovenia, 15–17 May 2007; WSEAS Press: Athens, Greece, 2007.

4. Zaharia, R.M.; Stancu, A.; Stoian, C.; Diaconu, M. Commercial Activity's Contribution to Sustainable Development by Social Responsability Actions: A Vision of SMEs. *Amfiteatru Econ.* **2010**, *XII*, 155–167.

5. Minciu, R.; Pădurean, M.; Popescu, D.; Hornoiu, R. Demand for Vacations/Travel in Protected Areas—Dimension of Tourists' Ecological Behavior. *Amfiteatru Econ.* **2012**, *XIV*, 99–113.

6. Rifkin, J. *La Tercera Revolución Industrial. Cómo el Poder Lateral Está Transformando la Energía, la Economía y el Mundo*; Paidós: Barcelona, Spain, 2011. (In Spanish)

7. Istrate, I.; Bran, F. *Economia Turismului și Mediul Înconjurător*; Editura Economică: Bucharest, Romania, 1996. (In Romanian)

8. Bran, F.; Simion, T.; Nistoreanu, P. *Ecoturism*; Editura Economică: Bucharest, Romania, 2000. (In Romanian)

9. Popescu, R.I.; Zamfir, A. Strategic Role of Ecotourism for Romania's Regional Development. In Proceedings of The 5th International Conference—The Scale of Globalization, Think Globally, Act Locally, Change Individually in the 21st Century, Ostrava, Czech Republic, 8–9 September 2011.

10. Stănciulescu, G. *Strategii si Politici Regionale de Dezvoltare Durabilă a Spaţiului Dunărean Provocări Pentru Turism*; Editura ASE: Bucharest, Romania, 2004. (In Romanian)

11. Minciu, R.; Popescu, D.; Pădurean, M.; Hornoiu, R.; Băltăreţu, A. Commercialization of Holidays in the Protected Natural Areas—Form of the Sustainable Development in Tourism.*Amfiteatru Econ.* **2010**, *XII*, 83–98.

12. Fernández, J.I.P.; Sánchez Rivero, M. Measuring Tourism Sustainability: Proposal for a Composite Index. *Tour. Econ.* **2009**, *15*, 277–296.

13. Kožić, I.; Mikulić, J. An Exploratory Assessment of Tourism Sustainability in Croatian Coastal Destinations. 2011. Available online: http://bib.irb.hr/datoteka/516656.Kozic_Mikulic_Tourism_Sustainability_Croatia_FEBWPS11-04.pdf (accessed on 25 May 2015).

14. White, V.; McCrum, G.; Blackstock, K.L.; Scott, A. Indicators and Sustainable Tourism: Literature Review. 2006. Available online: http://www.macaulay.ac.uk/ruralsustainability/LiteratureReview.pdf (accessed on 25 February 2015).

15. Corboş, R.A. Integration and Competition—Appropriate Approaches for Achieving Excellence in Management. *Bus. Excell. Manag.* **2011**, *1*, 67–73.

16. Jamieson, W.; Noble, A. A Manual for Sustainable Tourism Destination Management. 2000. Available online: http://www.gdrc.org/uem/eco-tour/etour-principles.html (accessed on 10 May 2015).

17. Zamfir, A.; Popescu, R.I. Promoting Renewable Energy—A Local Solution to Global Environmental Problems of Competitive Cities and Regions. In Proceedings of The 5th International Conference,

the Scale of Globalization, Ostrava, Czech Republic, 8–9 September 2011.

18. Alpopi, C.; Manole, C.; Colesca, S.E. Assessment of the Sustainable Urban Development Level through the Use of Indicators of Sustainability. *Theor. Empirical Res. Urban Manag.* **2011**, *6*, 78–87.

19. Bačlija, I. Reconceptualisation of Urban Management: Evidence from EU Cities. *Theor. Empirical Res. Urban Manag.* **2013**, *8*, 30–50.

20. Ţapardel, A.-C.; Alexe, F.-A. Strategic Directions for the Bucharest Strategy and City Brand. *Amfiteatru Econ.* **2012**, *XIV*, 720–737.

21. Jimenez Figueroa, Y.H.; Muro Pico, J.F. *Desarrollo Organizacional y Humano*; El Cid Editor: Santa Fe, Argentina, 2009; Available online: https://www.clubensayos.com/Informes-de-Libros/Resumen-Libro-Desarrollo-Organizacional-Y-Humano-JIMENEZ/886501.html (accessed on 25 May 2015). (In Spanish)

22. Argüden, Y. *Keys to Governance, Strategic Leadership for Quality of Life*; Palgrave Macmillan: Hampshire, UK, 2011.

23. Pantelescu, A.M.; Ioncică, M. Researching the Mass Media Role in Promoting and Selling Tourists' Packages in Romania. *Amfiteatru Econ.* **2012**, *XIV*, 802–810.

24. Stănciulescu, G.C.; State, O. Potential Interpretations of the Ecological Behaviour in Tourism. *Amfiteatru Econ.* **2012**, *XIV*, 114–128.

25. Huibin, X.; Marzuki, A.; Razak, A.A. Conceptualizing a Sustainable Development Model for Cultural Heritage Tourism In Asia. *Theor. Empirical Res. Urban Manag.* **2013**, *8*, 51–66.

26. Roberts, S.; Tribe, J. Sustainability Indicators for Small Tourism Enterprises—An Exploratory Perspective. *J. Sustain. Tour.* **2008**, *16*, 575–594.

27. Zamfir, A. *Management of Services within the Knowledge-Based Society*; Editura ASE: Bucharest, Romania, 2010.

28. Nedelea, A.; Bălan, A. E-tourism and Tourism Services Consumer Protection. *Amfiteatru Econ.* **2010**, *XII*, 492–503.

29. Minghetti, V.; Moretti, A.; Micelli, S. Reengineering the Museum's Role in the Tourism Value Chain: Towards an IT Business Model. *Inf. Technol. Tour.* **2002**, *4*, 131–143.

30. Polat, A. The Best City to Visit Travel Tournament 2013: Championship. 2013. Available online: http://foxnomad.com/2013/03/26/the-best-city-to-visit-travel-tournament-2013-championship/ (accessed on 15 June 2015).

31. Popescu, R.I.; Corboş, R.A. Strategic Options in the Construction of the Bucharest Brand through the Application Analysis of the Measuring Instruments for the Urban Brands. *Ann. Univ. Petrosani Econ.* **2010**, *10*, 267–278.

32. Zamfir, A.; Corboş, R.A. Development of Cultural Services within the Knowledge Economy: Case Study on the Romanian Museums. *Int. J. Arts Sci.* **2012**, *5*, 399–409.

33. Popescu, R.I.; Corboş, R.A. The Role of Festivals and Cultural Events in the Strategic Development of Cities. Recommendations for Urban Areas in Romania. *Inf. Econ.* **2012**, *16*, 19–28.

34. Popescu, R.I.; Corboş, R.A. The Role of Urban Brand Measuring Tools in Grounding the Strategic Actions for Promoting Cities. *J. Adm. Public Manag.* **2009**, *13*, 116–132.

9

CHAPTER

Perspectives on Cultural and Sustainable Rural Tourism in a Smart Region: The Case Study of Marmilla in Sardinia (Italy)

Chiara Garau

Department of Civil and Environmental Engineering and Architecture, DICAAR, University of Cagliari, via Santa Croce, 67, Cagliari 09124, Italy

ABSTRACT

This paper is being inserted into the current debate on the topic of sustainability, as it applies to rural tourism. In particular, it addresses the need to identify strategic actions that will enhance the dissemination of cultural resources to facilitate cultural planning. Balancing the dynamic tension that characterizes the relationship between tourism development and protection of the landscape is key to finalizing appropriate planning strategies and actions, especially in the context of marginal rural areas. In support of theoretical and methodological reflections pertinent to this relationship, this paper presents a case study of the region of Marmilla on Italy's island of Sardinia. The absence of both a "cultural planning" philosophy and a strategic approach to systemic and sustainable rural tourism in this country has been acknowledged. This paper concludes by discussing the results that emerged during the preparation of this case study, with respect to smart, sustainable, rural tourism development, while accepting the need for compromises between the force of globalization, nature, tourism, places, and people.

Keywords: cultural tourism; rural tourism; sustainable tourism; smart land use; cultural planning; Sardinia

1. INTRODUCTION

The phenomenon of rural tourism has recently assumed new significance, having risen gradually from a marginal to a widespread practice. Changes to the European countryside, increasing globalization, and growing competition between traditional locations and new destinations have raised tourist expectations, encouraging travel to places less well known, and recalling, for this reason, those tourist flows attracted by the authenticity of the experiences visitors are likely to have [1,2,3,4]. The pursuit of authenticity—consisting of cultural and social identities, traditions, memories, intangible connections, local peculiarities, and rural landscapes—has therefore led communities and local, national, and European governments [5] to respond to new tourism demands in more complex ways [6].

In Europe, this phenomenon has been consolidated over time, beginning in the late eighties. Since then, the literature has documented a wide range of theoretical paradigms aimed at interpreting these consequences, and the natural environment has seen the emergence of rural tourism. One of the first definitions of rural tourism was proposed by the European Commission in 1986: "Rural tourism is a broad concept that includes not only farm tourism or agritourism—accommodation provided by farmers—but all tourist activities in rural areas" [7]. Simonicca (1997) interpreted rural tourism as an alternative type of tourism with sustainable objectives [8]. Namely, socio-cultural or natural rural tourism environments represent alternatives to the places where tourists live, and they in turn try to experience positive and educational impacts from having visited them [9]. Daugstag (2008) defined the discovery of the rural land as "a refuge from urban life" [10]. That is, it represents an alternative to embracing globalization. Barke (2004) argued that rural tourism had developed in response to two factors. These were the "decline of traditional rural activities, principally agriculture and the consequent demographic changes, especially depopulation," and "the perceived need to diversify the [...] tourism product away from traditional mass beach tourism characteristic of the 1960s and 1970s" [11]. During that period, the phenomenon that Henry Lefebre called "the right to the city" [12] was tangible: the population had moved from rural areas to cities, mainly because the city provided the civitas, namely the social ties, functions, and services capable of providing an urban lifestyle. These conditions can occur in areas with small populations.

In the nineties, the population increased, as did the volume of information. Competitive territorial marketing strategies began to appear. They were aimed not only at improving the performance and increasing the attractiveness of the land, but also at maximizing its long-term economic benefits; they were also designed to ensure the sustainability of the territory, through proper management of tourist flows. Especially for rural areas, different forms of place-based enhancements such as eco-museums, farms, community maps (or parish maps), educational farms, and educational tours increased.

Corner and Swarbrooke (2004) emphasized the two main outcomes offered by rural tourism:*farm hosting*, in which country homes were used to provide hospitality, and *farm holidays*, during which the tourist was placed in rural areas and participated actively in the rural lifestyle that had been preserved as a primary agricultural activity [13].

Cawley and Gillmor (2008) argued that we could speak of rural tourism when there were strong links with the land's economic and productive activities, and three main features were in evidence: integration, sustainability, and endogeneity [14]. Schubert (2006), on the other hand, believed that rural tourism represented a key strategy for regional development [15]. Zhou (2014) asserted that rural tourism was by nature mainly "domestic, and positioned as a small-scale activity" [16], and Balestrieri (2005) highlighted the versatility of rural tourism, and suggested that it was for this reason that it was able to "play the role of engine of sustainable economic development" [17]. In sum, the rural landscape brought environmental sustainability together with different social, cultural, and economic components, duc to its different, content-rich, inter-connected, and integrated attractors, along with its strong anthropological characteristics, because it related to the culture and lifestyles of settled communities, and thereby enabled planning for tourism.

In view of global economic dynamics, planning and programming for the development of sustainable rural tourism represents one of the new challenges for strengthening and revitalizing lands that are otherwise not competitive. This paper begins by defining rural tourism as it has become established in Italy during the last ten years. Then, a case study of the region of Marmilla in Sardinia, Italy is presented. The author elaborates on a process for establishing a sustainable rural tourist destination, and describes the necessary cultural changes required of the different actors and tourist enterprises that would be involved in this process. It is assumed that Marmilla is a model of tourism development, and demonstrates best practices for rural cultural tourism, based on recent "smart region" paradigms [18,19], "Neogeography" [20,21], and the development of computing platforms that are increasingly more integrated and interactive [22]. Finally, sustainable rural tourism is discussed, using the study's findings. The paper concludes by recommending the development of smart rural-urban linkages, and demonstrating how even the planning and programming of rural tourism cannot escape comparisons with sustainability, which should be seen not as a constraint, but as a goal of contemporary management and place-based marketing.

2. SUSTAINABLE RURAL TOURISM IN ITALY: SOME REFERENCE DATA

Although tourism was one of Italy's most dynamic economic sectors in the recession period [23], the tourist presence has fluctuated in recent years. In fact, Italy's tourism sector data have shown a greater loss of market share than those recorded by other direct competitors such as Spain and France [24]. There are various reasons for this decrease: unstructured governance, fragmentary promotions abroad [25], a lack of engagement by the National Agency of Tourism [26], the Italian economic tourism sector being composed predominantly of small and medium-sized businesses [27], the lack of competitive tourism products, poor infrastructure, and a general lack of coordination at the political, technical, and operational decision-making levels [28]. In addition, there are problematic aspects related to the use of digital platforms, such as the lack of a digital tourism strategy, and the insufficient use of digital sales channels and applications for smartphones and tablets. To address these weaknesses, the *Strategic Plan for the Development of Tourism in Italy* (*Piano strategico per lo sviluppo del turismo in Italia* (2013)) identifies targeted and specific actions that are to be undertaken: the construction of a laboratory for e-tourism; the development of mobile applications specifically targeted at foreign tourists, who know how to integrate logistical information with cultural resources (such as museums and exhibitions); the definition of roles and responsibilities; and the coordination of governance arrangements, not only between the state and the regions/provinces, but also between provinces and municipalities [29]. Service integration, mobile applications, and the coordination of decision making at all levels are key aspects of national tourism innovations, and they need to be relaunched in a smart way.

These factors have also been of interest in rural areas, where they have produced different results. In these areas, in fact, there have been positive trends, especially from 2003–2013. Although there have been no ad hoc surveys on the phenomenon, and although there are objective difficulties involved in identifying a destination's unique rural character [30], this trend toward a rural tourism sector is apparent from the consistent growth of agritourism farms.

As shown in Table 1, their number rose from 13,019 in 2003, to 20,897 in 2013, thereby increasing by 60.5%. Firms dedicated to other agritourism activities have acquired an important role. They have shown increases of 47.9% and 62.7% for tasting and other activities respectively. In the same period, the beds available have increased by 94,738, and the seating capacity has increased by 157,615 [31]. The major "boom" in 2013, however, concerned activities related to "educational farms", which, compared to 2010—the year they began to appear—have increased by 15%. While central Italy offers agritourism farms with greater diversification (Table 2), in the period between 2012 and 2013, agritourism farms grew more significantly in the north at 6.1%, relative to those in the center and on the islands; in the south, a decline of 5.8% was observed (Table 3) [31]. The regions of Tuscany and Trentino Alto Adige have the most agritourism farms (in 2013 they exceeded 3500 units). Agricultural hospitality is more rooted there for historical reasons. Agritourism farms are also widespread in the regions of Lombardy, Veneto, Umbria, Piedmont, and Emilia Romagna (with over 1000 companies), and, finally, Lazio, Marche, and Sardinia.

Although the data described above tend to underestimate the phenomenon [32], they confirm that rural tourism is a reality with its own identity, and a well-characterized question. Interest in this sector is destined to grow, probably, as mentioned in the previous paragraph, due to evolving tourism preferences and demands. A large part of rural Italian tourism is still very much influenced by several factors, including the fragmentation and lack of cohesion of the valorization choices in an extensive rural area; local communities' disinclination to accept change, being too weak to assume the stresses that accompany innovative tourism; and the fairly limited capacities of existing hospitality structures.

Linking rural tourism with cultural tourism can lead to the integration and diversification of tourism opportunities, and translate the tourist destination as a whole into a competitive destination. In fact, the combination of cultural and rural offerings are able to support political, institutional, or proactive choices for new tourism products, and may represent a unique experience [3,4], not only for the tourist in search of authenticity, but also for all other stakeholders (the local communities, local administrators, institutions, non-profit organizations, and business employees) involved in the territory. Rurality, widespread cultural heritage, and tourism therefore comprise a complex relationship that imposes on territories and local actors the expectation that they will clarify their tourism potential, and the related processes already in place. In this way, the rural area is no longer a "product", but is rather a place where sustainable place-based integrated development processes are activated [33].

Sardinia was chosen for this study for several reasons. First, this island's rural tourism processes have never before been considered as offering an alternative, more consolidated development model relative to the seaside model. Rural tourism has always been regarded as secondary to seaside tourism. The development of a rural development model, however, could facilitate the attenuation of Sardinia's imbalances by countering the seasonal ebbs and flows characteristic of seaside tourism. Secondly, Sardinia ranks eleventh in the total number of Italian farms by region, and represents the median number of farms (Table 3). For these reasons, Sardinia is considered to be representative of the rural Italian reality. This case study of the natural region of Marmilla in Sardinia also addresses questions associated with integrating the cultural, natural, and rural heritage. Even though the Marmilla region has this significant potential, a strategic process for tourism planning has not yet been created. In this paper, based on the following concepts: tourism, rurality, and cultural heritage, the author defines the conditions needed to initiate a process of reflection that could lead to defining a strategic tourism planning process.

3. GOVERNANCE AND MANAGEMENT OF MARMILLA'S PLACE-BASED HERITAGE FROM A SUSTAINABILITY PERSPECTIVE

Based on the preceding information, the question that arises is how to initiate a sustainable tourism approach for an agricultural destination that does not yet have a strategic planning and management approach for its tourism sector. We begin this research by analyzing Marmilla's rural and natural landscape, as well as its cultural and historic heritage. Doing so allows us to understand the extent and value of Marmilla's cultural and environmental aspects. Next, socio-demographic and production supply and demand dynamics are analyzed, to help delineate an appropriate strategic policy for place-based development and tourism

Table 1. Typologies of agritourism farms in Italy (ISTAT Data, 2014).

Typologies of agritourism farms	2003	2004	2005	2006	2007	2008	2009	2010	2011	2012	2013	Difference 2003–2013 Absolute Number (Abs.)	%
Accommodation													
- Firms	10,767	11,575	12,593	13,854	14,822	15,334	15,681	16,504	16,759	16,906	17,102	6335	58.8
- Number of beds	130,195	140,685	150,856	167,087	179,985	189,013	193,480	206,145	210,747	217,946	224,933	94,738	72.8
- Picnic areas	4540	5386	5826	6935	7055	7320	7785	8759	9113	8363	8180	3640	80.2
Food & beverage													
- Firms	6193	6833	7201	7898	8516	8928	9335	9914	10,033	10,144	10,514	4321	69.8
- Seating capacity	249,342	266,654	277,866	258,003	322,145	337,385	365,943	385,470	385,075	397,175	406,957	157,615	63.2
Tasting													
- Firms	2426	2737	2542	2664	3224	3304	3400	3836	3876	3449	3588	1162	47.9
Other Activities													
- Firms	7436	8240	8755	9643	9715	10,354	10,583	11,421	11,785	11,982	12,096	4660	62.7
of which													
- Horse Riding	1364	1494	1478	1557	1559	1615	1548	1638	1662	1489	1230	−134	-9.8
- Escursionism	2452	2692	2981	3131	2879	3140	3071	3190	3233	3324	3124	672	27.4
- Naturalistic Obs.	224	265	575	517	558	607	623	784	891	932	972	748	333.9
- Trekking	1350	1463	1426	1465	1629	1657	1674	1950	1949	1821	1717	367	27.2
- Mountain Bike	2101	2422	2258	2311	2347	2398	2309	2800	2794	2785	2851	750	35.7
- Educational Farms	-	-	-	-	-	-	-	752	1122	1251	1176	1176	-
- Courses	693	812	942	1025	1256	1407	974	1967	1878	2009	1770	1077	155.4
- Sports	2927	3006	3474	3682	3758	4203	4168	4152	4141	5058	5088	2161	73.8
- Various	3786	4003	4288	5043	5395	5616	5994	6312	6737	4917	6033	2247	59.4
Agritourism Farms													
- Total Firms	**13,019**	**14,017**	**15,327**	**16,765**	**17,720**	**18,480**	**19,019**	**19,973**	**20,413**	**20,474**	**20,897**	**7878**	**60.5**

Table 2. Agritourism farms in Italy by type of activity and region (ISTAT Data, 2014).

Regions Geographical distributions	Food & beverages				Tasting				Other Activities			
	2012	2013	Difference		2012	2013	Difference		2012	2013	Difference	
			Abs.	%			Abs.	%			Abs.	%
Piedmont	753	790	37	4.9	589	616	27	4.6	902	925	23	2.5
Aosta Valley	45	36	−9	−20.0	35	9	−26	−74.3	10	9	−1	−10.0
Lombardy	1019	1060	41	4.0	116	144	28	24.1	673	722	49	7.3
Trentino-Alto Adige	577	625	48	8.3	100	108	8	8.0	1311	1348	37	2.8
Bolzano-Bozen	430	470	40	9.3	-	-	-	-	1255	1,292	37	2.9
Trento	147	155	8	5.4	100	108	8	8.0	56	56	-	-
Veneto	756	782	26	3.4	601	641	40	6.7	511	524	13	2.5
Friuli-Venezia Giulia	447	454	7	1.6	10	13	3	30.0	229	240	11	4.8
Liguria	281	353	72	25.6	-	40	40	-	336	287	−49	−14.6
Emilia-Romagna	797	834	37	4.6	-	-	-	-	874	739	−135	−15.4
Tuscany	1131	1232	101	8.9	577	515	−62	−10.7	2925	3141	216	7.4
Umbria	405	409	4	1.0	227	237	10	4.4	1108	1120	12	1.1
Marche	414	447	33	8.0	380	420	40	10.5	306	234	−72	−23.5
Lazio	551	596	45	8.2	133	162	29	21.8	552	571	19	3.4
Abruzzo	436	410	−26	−6.0	73	56	−17	−23.3	467	377	−90	−19.3
Molise	86	86	-	-	50	50	-	-	54	54	-	-
Campania	352	396	44	12.5	136	151	15	11.0	287	330	43	15.0
Puglia	271	222	−49	−18.1	146	138	−8	−5.5	231	303	72	31.2
Basilicata	98	78	−20	−20.4	40	20	−20	−50.0	104	54	−50	−48.1
Calabria	569	542	−27	−4.7	50	49	−1	−2.0	503	472	−31	−6.2
Sicily	473	493	20	4.2	186	219	33	17.7	514	550	36	7.0
Sardinia	683	669	−14	−2.0	-	-	-	-	85	96	11	12.9
ITALY	10,144	10,514	370	3.6	3449	3588	139	4.0	11,982	12,096	114	1.0
Northern Italy	4675	4934	307	6.6	1451	1571	128	8.8	4846	4794	−15	−0.3
Central Italy	2501	2684	183	7.3	1317	1334	17	1.3	4891	5066	175	3.6
Mezzogiorno	2968	2896	−27	−0.9	681	683	31	4.6	2245	2236	10	0.4
South Italy	1812	1734	−78	−4.3	495	464	−31	−6.3	1646	1590	−56	−3.4
Islands	1156	1162	6	0.5	186	219	33	17.7	599	646	47	7.8

Table 3. Agritourism farms in Italy by region (ISTAT Data, 2014).

Regions Geographical distributions	2012	2013	Difference		Regions Geographical distributions	2012	2013	Difference	
			Abs.	%				Abs.	%
Piedmont	1164	1220	56	4.8	Umbria	1262	1280	18	1.4
Aosta Valley	54	53	−1	−1.9	Marche	788	880	92	11.7
Lombardy	1415	1521	106	7.5	Lazio	841	884	43	5.1
Trentino-Alto Adige	3391	3506	115	3.4	Abruzzo	774	653	−121	−15.6
Bolzano-Bozen	2996	3098	102	3.4	Molise	104	104	-	-
Trento	395	408	13	3.3	Campania	407	458	51	12.5
Veneto	1376	1449	73	5.3	Puglia	355	353	−2	−0.6
Friuli-Venezia Giulia	588	614	26	4.4	Basilicata	145	112	−33	−22.8
Liguria	543	567	24	4.4	Calabria	610	577	−33	−5.4
Emilia-Romagna	1036	1106	70	6.8	Sicily	602	633	31	5.1
Tuscany	4185	4108	−77	−1.8	Sardinia	834	819	−15	−1.8
ITALY	**20,474**	**20,897**	**423**	**2.1**	**Mezzogiorno**	**3831**	**3709**	**−79**	**−2.1**
Northern Italy	**9567**	**10,036**	**584**	**6.1**	**South Italy**	**2395**	**2257**	**−138**	**−5.8**
Central Italy	**7076**	**7152**	**76**	**1.1**	**Islands**	**1436**	**1452**	**16**	**1.1**

planning. Finally, using different scales, the application of direct and indirect governance instruments as they might affect the development of tourism and environmental protection in the Marmilla context are analyzed. These analyses reveal that a successful strategic planning and management approach for Marmilla depends primarily on social capital. Indeed, awareness of local administrators regarding their roles and the potential of a strategic approach to systemic and sustainable rural tourism will provide the conditions needed to improve the attractiveness (and therefore also the place-branding) of the cultural and natural heritage, and make a significant contribution in terms of political cohesion, identity and local development. In this regard, it has been observed that the role of technology can be crucial to the debate on governance for sustainable tourism in rural areas. Indeed, Go, Della Lucia, Trunfio, and Martini (2014) state that there is the need for a link between "two broad knowledge domains, ICTs and place branding [of rural contexts], often isolated from one another" [34]. Tying these two aspects does not delegate to rural areas the characteristic of "sustainable place", but offers them the opportunity to have a voice, visibility and place-based development in the era of globalization [34]. In addition, it captures the need to overcome certain structural limitations, creating, for example, a "system of rural areas": especially in Marmilla, a recent study showed that four of its 18 municipalities (Genuri, Tuili Turri, and Ussarramanna) are at risk of demographic desertification [35].

For this reason, in this paper, with the help of new technologies, the most relevant factors to emerge from previous analyses are identified, collected, and "filtered", using a logical framework, and hypothesizing a scenario aimed at two different but parallel objectives: protecting the natural and cultural heritage, and advancing sustainable tourism development in the case study region of Marmilla, while accepting the need for compromises between the forces of globalization, nature, tourism, places, and people.

3.1. Case Study of Marmilla

The region of Marmilla is located in the southern-central part of Sardinia, Italy, in the province of the Middle Campidano (Figure 1). It covers 415 square kilometers, and had a population of 25,619 in 2014 [36].

Marmilla's morphology is mainly devoted to agriculture, supported by a system of small settlements. It comprises 18 municipalities [37] and, as shown in Table 4, is characterized by features typical of southern Italy's inland areas: a low population density, weak economic structures, negative long term demographic trends (Figure 2 and Table 5), and an aging population [38,39,40]. The social and demographic situation is not uniform, and some of the 18 municipalities do not exceed 300 inhabitants (such as Setzu and Las Plassas with 151 and 240 inhabitants respectively), while others have more than 8500 inhabitants (Sanluri). The territory is characterized by a strong reliance on commuting, especially closer to major urban centers such as Cagliari, Oristano, and Sassari.

Figure 1. The natural region of Marmilla.

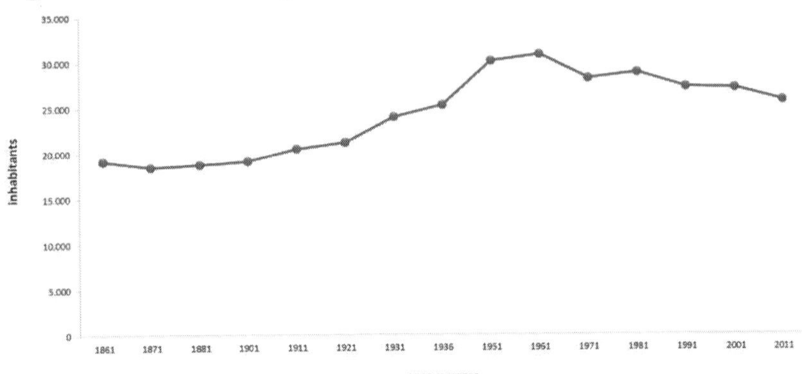

Figure 2. Marmilla's demographic trend from 1861 to 2011 [41].

Table 4. Socio-demographic and economic dynamics in Marmilla [42].

Geographical distributions	Area (km square)	Population density (2014)	Population 2014 (inhabitants)	Aging index 2014	Disposable Income Per capita 2013 (Income—Taxes [fiscal levy])
Barumini	26.57	48.78	1296	261.4	€ 11,986.00
Collinas	20.79	41.41	861	292.9	€ 12,507.00
Furtei	26.12	64.20	1677	204.8	€ 11,117.00
Genuri	7.55	45.03	340	463.6	€ 12,855.00
Gesturi	46.87	27.29	1279	319.4	€ 10,596.00
Las Plassas	11.14	22.08	246	255.6	€ 10,596.00
Lunamatrona	20.57	85.03	1749	277.7	€ 13,028.00
Pauli Arbarei	15.12	42.72	646	247.7	€ 11,465.00
Sanluri	83.78	101.81	8530	174.5	€ 14,418.00
Segariu	16.69	74.18	1238	206.9	€ 10,596.00
Setzu	7.82	19.31	151	273.7	€ 14,071.00
Siddi	11.02	61.43	677	321.2	€ 11,291.00
Tuili	24.5	42.86	1050	375.5	€ 11,986.00
Turri	9.64	46.37	447	442.1	€ 11,986.00
Ussaramanna	9.75	57.23	558	305.0	€ 13,376.00
Villamar	38.64	72.93	2818	174.0	€ 11,465.00
Villanovaforru	10.97	59.16	649	272.3	€ 11,986.00
Villanovafranca	27.46	51.24	1407	283.1	€ 10,944.00
Marmilla	415	61.73	25,619	226.5	€ 12,014.94
Sardinia	24,090	69.07	1,663,859	174.4	€ 13,871.00
Italy	301,338	201.71	60,782,668	154.1	€ 17,038.20

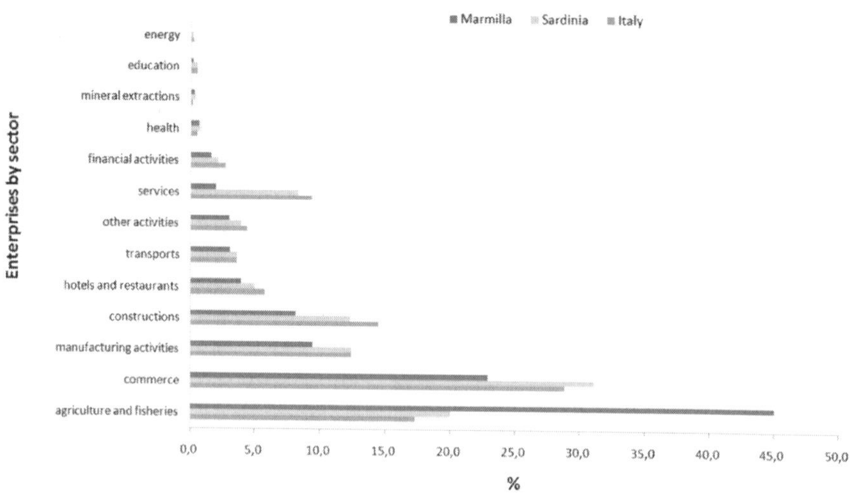

Figure 3. Segmentation by percentage of enterprises (year 2010) [43].

Table 5. Marmilla's demographic trend from 1861 to 2011 [41].

Geographical distributions	1861	1871	1881	1901	1911	1921	1931	1936	1951	1961	1971	1981	1991	2001	2011
Barumini	1214	1187	1221	1118	1179	1335	1445	1431	1685	1729	1647	1516	1423	1413	1310
Collinas	976	1012	1072	1033	1088	1065	1040	1091	1206	1213	1129	1145	1076	1014	885
Furtei	1030	915	981	1057	1118	1179	1280	1422	1728	1846	1788	1830	1729	1723	1674
Genuri	359	400	434	383	440	446	535	575	654	706	567	518	444	386	345
Gesturi	1660	1457	1430	1431	1507	1455	1643	1709	1827	1801	1567	1515	1438	1430	1280
Las Plassas	486	459	429	397	454	500	587	502	566	632	379	298	291	269	257
Lunamatrona	968	1018	1104	1148	1299	1278	1467	1640	1948	2017	1850	1896	1896	1858	1783
Pauli Arbarei	433	424	409	401	477	530	656	676	801	797	787	778	692	720	651
Sanluri	4199	4177	4177	4403	4593	4786	5449	5721	7555	7595	7402	8305	7912	8519	8460
Segariu	700	588	647	661	732	750	899	989	1308	1441	1409	1432	1320	1358	1277
Setzu	302	330	276	240	292	267	304	325	278	278	223	223	184	166	144
Siddi	608	615	603	636	631	802	869	871	987	1121	990	903	869	799	696
Tuili	1334	1286	1242	1320	1330	1302	1478	1613	1713	1591	1348	1347	1263	1185	1062
Turri	454	488	511	503	490	487	575	602	729	734	633	597	572	533	442
Ussaramanna	623	640	609	586	677	730	790	863	920	963	835	714	656	611	556
Villamar	1948	1825	1903	2047	2250	2220	2675	2876	3301	3369	3057	3196	3147	2960	2872
Villanovaforru	506	517	593	615	655	709	741	770	905	931	846	789	733	700	681
Villanovafranca	1356	1166	1121	1189	1286	1369	1577	1633	2055	2117	1759	1871	1621	1491	1433
Marmilla	**19,156**	**18,504**	**18,762**	**19,168**	**20,498**	**21,210**	**24,010**	**25,309**	**30,166**	**30,881**	**28,216**	**28,873**	**27,266**	**27,135**	**25,808**
Sardinia	**609,015**	**636,413**	**680,450**	**795,793**	**868,181**	**885,467**	**983,760**	**1,034,206**	**1,276,023**	**1,419,362**	**1,473,800**	**1,594,175**	**1.648,248**	**1,631,880**	**1,639,362**

Extensive territories with high environmental quality and low population levels contribute to conserving the land over time, and maintaining its rurality (Figure 3). On the other hand, the region lacks a socially young and dynamic cohort. These issues have become structural problems for place-based development in Marmilla, and have led the Marmilla region to be perceived as being on the margins of Sardinia's economy.

In this context, the development of tourism could limit the impacts of the critical weaknesses described above, by offering new land-based job opportunities to members of the local community, thereby benefitting the entire economy and society. Natural, cultural, and environmental attractions abound: the uniqueness of the landscape, the historical value of the small urban centers, and the wealth of traditions linked to the agricultural economy, united by the presence of many archaeological and nuraghic remains. Among the latter, the Nuraghe of Barumini (included in the United Nations Educational, Scientific and Cultural Organization's [UNESCO] list of World Heritage Sites in 1997) is a prime destination in the region of Marmilla, and its draw benefits the whole island of Sardinia. However, tourist groups that visit the Nuraghe often do not consider the island's many other attractions, and so the Nuraghe do not have an appreciable economic impact on the extended area, in terms of employment or stimulating commercial activities.

The municipalities in Marmilla have tried in recent decades to adopt a systematic approach to building a local shared identity, by establishing numerous associations [44]. Government administrators have tried to diversify tourism sites by proposing—unfortunately only on paper—the integration of natural attractions (such as the Geobotanical Park, the Museum of the Territory, and the Jars of Siddi, Collinas, Tuili, and Setzu), with archaeological sites (such as the various archaeological museums and the many areas in which Nuraghes, sacred wells, and the Tombs of Giants are present), and with folkloristic, ethnographic, demo-ethno-anthropological, and historical points of interest (small museums, Romanesque churches, altarpieces from the sixteenth century, cultural centers, festivals and village traditions) [45].

In other words, local administrators have not managed to effectively organize a real "team" that could assume responsibility for the formulation of a strategic plan for place-based tourism, using place-based logic to enrich the local community's purely parochial visions. Maybe this is because (temporary) local ambitions promoted place-based identities that did not represent the local community in the best possible light: the area of Marmilla, as mentioned before, has historically been dedicated to agriculture and pastoralism. Not including these aspects in the range of tourist offerings has definitely compromised its place-based, economic, and social development. It is clear that the context for development will not be mass tourism, but rather it will have to consider the interrelationships between sustainability, the local culture, the rural environment, the community, and tourism. The peculiarities of the rural land must remain unchanged, without becoming part of a vicious cycle in which they are sacrificed for economic development.

Other Italian regions (such as Umbria and Marche) have managed to create place-based brands for some areas, and have therefore been able to offer competitive tourism that is linked to the natural characteristics of the agricultural and mountainous sectors.

The typical quality of life in rural areas, the reasonable cost of living, and its central location in Sardinia could combine to make Marmilla a strategic tourist destination, if positive hospitality, commercial, and tourism sector processes are actually implemented.

3.2. Development of the Tourism Sector in Marmilla

According to published literature, sustainability has environmental, social, and economic dimensions [46,47]. Tourism development is thus sustainable only if has been strategically planned to achieve goals that manifest their effects over the long term. For this reason, Marmilla's strategic programming framework was analyzed at different scales, considering various government instruments such as the Rural Development Programme (RDP) 2014–2020 [48], the Regional Development Plan (RDP) 2014–2019 [49], urban master plans (for all eighteen municipalities), and the spatial strategic plan of Sanluri (the only municipality in Marmilla using this urban tool).

This comparative analysis facilitated the development of a vision for strategic and planned processes in the case study area. It also contributed to an understanding of how best to enhance integrated planning policies for a public/private initiative, and, most importantly, how to requalify rural lands for landscape protection, and for cultural and sustainable rural development (Table 6).

The principles summarized in Table 6 provide insight into how best to orient local communities in Marmilla to these processes. However, the most important innovations for triggering development in this rural area are of two types. These are the region's social capital, and the latest technologies capabilities for raising awareness of the area's rural realities worldwide, and making them attractive for tourists interested in this niche sector.

The local community plays a crucial role in providing sustainable social capital, and must therefore participate in the development process for the long term [50]. Entrepreneurs will be able to increase the area's place-based competitiveness with their innovations [51]. Go, Della Lucia, Trunfio, and Martini (2014) emphasize how important it is to consider social capital under the structural [52], cognitive [53], and relational [54] dimensional perspectives.

Table 6. Evaluation criteria for redeveloping the area of Marmilla.

Protection and Cultural Landscape	Development of Sustainable Tourism
ENVIRONMENTAL SUSTAINABILITY; Preservation of landscapes, habitats, and ecosystems; Promotion of the use of renewable resources; Introduction or improvement of environmental management systems; Protection of the main territorial vocations	LOCAL ECONOMIC DEVELOPMENT; Encouraging the development of local firms and businesses; Encouraging the formation of employment aimed at sustainability of the tourism sector; Encouraging public and private partnerships; Promoting the construction or renovation of buildings for rural tourism, sustainable in the long term, despite changing political mandates
ECONOMIC SUSTAINABILITY; Development of landscape quality recognized by international bodies (UNESCO, *etc.*); Development of a market for local goods and sustainable services; Encouraging investments in innovative, environmentally friendly technologies	SUSTAINABLE ENVIRONMENTAL DEVELOPMENT; Encouraging initiatives dedicated to diversification of tourism and to the redistribution of tourist flows; Protecting and promoting the cultural and historical heritage; Encouraging the demand for and the achievement of environmental certification in the tourism sector
SOCIAL SUSTAINABILITY; Triggering processes of awareness of local tangible and intangible goods; Activating processes to raise awareness of the topic of sustainable development, protection, and enhancement of cultural heritage and public spaces; Improving the participatory practices; Improving the network of relationships between various stakeholders	SOCIAL DEVELOPMENT; Developing the integration between different policy sectors; Construction of participatory practices aimed at promoting the latest information technology tools for tourism purposes; Developing a place-based approach in which the local community is integrated with tourists; Promoting opportunities that induce residents to identify unique regional elements

For these reasons, particularly in Marmilla there is a need to maintain the area's traditional and historical vocational aspects [55]. Tourists' destination choices in turn may lead to a change of original vocation and increasing competitiveness—but only temporarily (in relation to the structural perspective). It is also important that Marmilla's administrators, including legislators, coordinators, and promoters, start the development process, while conducting their different roles and responsibilities, in line with the dimensional perspective [56]. And finally, the local population's "internal" visions and tourists' "external" visions for a rural place—that constitute the relational dimension of social capital—can help to define a common agenda. In so doing, they will help to identify goals and objectives for developing and promoting the area. For that reason, it is also important to use today's communication and marketing technologies to promote and embed knowledge of the attractiveness of the local area in national and international tourist markets. With regard to the latter point, it will be helpful to identify incentives associated with particular ecological brands recognized by the World Trade Organization (WTO) in the international context such as *Ecolabel*, or *Orange Flag*; *Organic Farms*; Associazione Borghi Autentici d'Italia (*The Association of Italian Authentic Villages*); and *I Borghi più Belli d'Italia* (*The most Beautiful Villages of Italy*), promoted by the National Association of Italian Municipalities (ANCI) in the national context.

There have been some sporadic adhesions of this type in Marmilla's municipalities. Collinas joined *The Association of Italian Authentic Villages*. The farm *su Massaiu* of Turri, which represents an example of excellence at the local level, has joined *Ecolabel*. In fact, su Massaiu produces a range of local melons, saffron, and a type of grain called *Senatore Cappelli*. Orchards, olive groves, almonds, medicinal herbs, legumes, and vegetables in general are also cultivated. The farms serve these products to customers or sell them. The benefits of owning such brands are significant. External communications are strengthened by combining the brand's organization (or brand's destination) with the brand's certification, and a virtual path is taken to manage the activities necessary for achieving and maintaining such recognition.

Regarding the second point, using the latest ubiquitous technology to foster, external visibility can do much to leverage development of the Marmilla area, and open the way for innovation, without sacrificing the local cultural and environmental identity. This hypothesis, currently in the concept stage, is further discussed in the following section.

4. TOURISM AND NEW MEDIA, A POSSIBLE COMBINATION IN RURAL CONTEXTS?

In recent years, global organizations—including the International Council on Monuments and Sites (ICOMOS), UNESCO, and the WTO—have promoted cultural tourism [57], using new technologies [58]. Innovative economic development projects aimed at creating "Smart

Regions" [18,19], the application of tools associated with "Neogeography" [20,21], and the development of more integrated and interactive computing platforms [22], have led to many experiments, including some in the field of tourism. Over the last decade, the main innovations in this field have included the creation of specific Internet portals and smart cards, fostering a locale's cultural heritage, and, finally, the diffusion of mobile tourism applications. Technology has allowed the formation of more dynamic and "immersive" relationships between tourists, the area's cultural heritage, and the region [58,59].

Ever-present technology has impacted consumers' demands for vacations, as well as the travel and tour packages being offered by suppliers. It has also enabled the diffusion of tourism information in a manner that has ensured maximum customization and user interaction, in contrast to the influence of traditional paper maps or guidebooks on these activities [59].

With continuing innovations, the information communication technologies (ICT), social networks, and mobile applications can facilitate the integration of tourism products in different moments of the consumption process. They have opened the doors to online marketing, strategic management, and marketing information systems (MkIS). They have changed how tourists search for information on different places and evaluate their alternatives. Service providers can now conduct surveys on tourism behaviors before and after the tourists select a desired destination.

However, in a rural and marginal area such as Marmilla, which is firmly anchored in tradition, the use of technological tools will have to be well understood and supported by the local community if they are to be accepted. These tools appear to be effective when they are accompanied by continuous training for the local communities, thereby creating a self-renewing and reproducing process over time.

In Marmilla then, cultural, rural, and natural excellence needs to be identified, communicated (through various tools, from online to offline dashboards), and systematized, and then used as part of a strategic approach for tourism governance. Porrello (2006) also included cultural planning as an essential part of the overall perspective [60]. In Italy, there are sporadic cases in which applications downloadable on smartphones or tablets have been used to promote tourism in rural areas. Of Italian regions, Tuscany has been the most active [61], and even if their communications through digital apps are not organized in an interactive way, they look very similar to website consultations. This mode of technological fruition in rural places is still not widespread in Sardinia. For this reason, the organized and attractive simulation of rural tourism paths that have been realized using the Garau's model [19,58,59] is proposed.

The implementation of rural pathways has led to thinking of whole, place-based involvement, as part of a strategic logic network that will enhance the resources of individual municipalities for the benefit of the whole region of Marmilla. On the basis of the analyses conducted here (Section 3.1), and considering the need to improve social and technological capital (Section 3.2), the author has identified a number of preliminary strategic actions that are linked to the enhancement of the area's cultural and rural heritage, and to the development of sustainable tourism. These actions can be considered as a set of best practices, and a place for local administrators to start, before beginning work on a cultural planning tourism development model.

The recommended actions are listed below:

(1) There is an overarching need to promote sustainable tourism across Marmilla, and not only in the main polarity (for instance what today includes the Nuraghe of Barumini), by highlighting the region's rural and cultural aspects and creating an integrated quality. These aspects include local museums, exhibitions, cultural events, and the interrelation of individual museums associated with rural structures, and interactive educational farms and/or guided tours as part of ongoing rural processes—such as harvesting grain and saffron.

(2) A second required action is developing the skills needed to begin proper planning and programming for rural tourism, through a centralized control, in which experts do not have local interests and are capable of supporting the often fragmentary and conflicting dynamics typical of rural areas. This can be achieved by involving experts who are external to the region.

(3) Next, it will be necessary to enhance local places' competitiveness (for example, by returning to traditional ways of promoting eco-museums, permanent, temporary, or itinerant events, exhibitions, and installations) and entrepreneurial tourism (through prizes or incentives to operators and companies that are distinguished by the quality offered, or through actions that stimulate an increase in the number of beds offered, while basing projects on ethical and sustainable development models).

(4) A fourth action will involve integrating agricultural activities with services that are compatible with tourism activities, by proposing shop windows that display local products.

(5) The visibility of local attractiveness should be unified, by creating a place-based promotion of the entire region of Marmilla (from the

establishment of a centralized structure that encourages the formation of networks of rural enterprises, to joint agreement on a unique logo for the area and/or for place-based marketing).

(6) Demand loyalty should be strengthened, through a series of actions aimed at enhancing internal communications among municipalities, and external communications between Marmilla and the rest of the world (for example, launching marketing actions on specific segments; promoting new tourism packages; conducting surveys to understand visitors' motivations for coming, enhancing the interest and attractiveness of new offers; improving web marketing actions; promoting seasonal offers; and encouraging the movement of visitors from established attractions to previously unknown places of interest).

(7) Finally, a commitment to sustainability should be used as the parameter for planning tourism in the area. On this point, it is essential that common goals and partnerships among the parties involved be identified, to foster understanding and to periodically update the processes, strategies, and planning associated with tourism development.

These actions, which are summarized in Figure 4, can be merged with the definition of rural pathways for Marmilla's place-based tourism redevelopment. Furthermore, the author has identified ways to enhance the quality and integration of local resources, while on the one hand respecting their historical identity, and on the other hand trying to better understand cultural offerings appropriate for rural tourism. This was achieved by reviewing existing literature on the area's cultural heritage and rural offerings [62,63,64,65,66], and from informal discussions with residents and administrators.

The notion of using different pathways, currently in the concept stage, touches the main points of interest (POIs) with a series of customizable stops that are linked to the availability of opportunities to taste local products, visit the historical, architectural, and archaeological museums representative of Marmilla's widespread heritage, and to visit landscaping, natural, and environmental landmarks and permanent agritourism facilities or educational farms across the region.

The proposal for the pathways is premised on the development of an online platform, downloadable from a smartphone or tablet, but supported by printed materials, and capable of being manipulated by the local community. Media dissemination of mobile technologies could attract people who would not otherwise be aware of Marmilla's potential.

The proposed paths would thus create smart rural-urban linkages, and demonstrate how the planning of and programming for rural tourism cannot escape a comparison with sustainability. Furthermore, in a strategic planning process for the tourism sector, the development of more attractive local organizations could lead to advantages for the whole island, by creating tourism districts for example, and increasing, consequently, the attractiveness of the regional offer [67].

Figure 4. Rural pathways in Marmilla.

5. CONCLUSIONS

The challenge that has become apparent by analyzing the Marmilla case study is the need to pursue the goals of place-based improvements, competitiveness, and sustainability in the longer term for this land-locked and marginal area of Sardinia. This perspective—extensible to other national or international rural contexts—is aimed at increasing the number of rural tourist destinations available today, considering the context of Marmilla as a smart tourist destination, and accepting necessary "compromises" between globalization, nature, tourism, places, and people.

As has been discussed, tourism has been used to promote the development of the region of Marmilla by promoting its strengths (the Nuraghe of Barumini), and making improvements that, until now, have been linked to less "traditional" tourist flows. In fact, the Nuraghe of Barumini is a well-known site in the tourism market, but that site alone is unable to trigger new endogenous development across Sardinia.

The integration of the various factors described in this work—such as rural lands, the cultural, archaeological, and natural heritage, local and traditional food production, and networks between all the actors—can constitute a winning tourism offering for tackling the socio-economic issues of rural areas in general, and of the agricultural sector in particular. Tourism structured in this way can trigger collaborative processes among Marmilla's different administrative offices, and thereby initiate actions that will precipitate endogenous development. Namely, they can create the conditions necessary for future developments that include "aggregating tourist offerings", by integrating or diversifying their offerings on the basis of new and different targets such as school tourism, tourism of the third age, religious tourism, and folkloristic tourism.

Strong values consistent with environmental sustainability, and proper diversification of the social fabric and local business community are keys to achieving an essential balance in the case study presented. This vital process must be based on a centralized control that can interface with all the actors, promote their potentialities, and mediate any conflicts. Fragmentary and parochial visions typical of rural areas inevitably lead to heterogeneous approaches to planning and programming tourism products [68]. This difficulty has to be faced and overcome in order to develop tourism products in a time of globalization, when tourists can experience all offerings for a single destination as unique, and cannot be focused on one firm or a single cultural or archaeological product. The image of a destination that emphasizes its vocations and its local cultural heritage is competitive and attractive in itself. It is important to reach the right target, by focusing less on marketing individual products, and emphasizing the uniqueness of the emotional experiences offered.

New forms of communication such as the Internet, social networks, and applications downloadable on smartphones or tablets, such as the one presented in this paper, are essential to improving the visibility and enhancing the promotion of tourism opportunities available in rural Marmilla. However, in this area, where some aspects of a digital divide remain, this form of communication has to be accompanied by formal and informal offline networks. Both modalities of communication are needed to promote the area in the long term, without relying on occasional advertising campaigns. Nevertheless, we must not forget that the successful integration of rural and cultural tourism can be achieved if there is agreement among the various stakeholders—including the local and regional administrators and the local community—to coordinate and encourage the development of local resources. They can, in fact, intervene and help the local tourism sector of Marmilla evolve for the better.

CONFLICTS OF INTEREST

The author declares no conflict of interest.

REFERENCES AND NOTES

1. McCannell, D. *The Tourist: A New Theory of the Leisure Class*; University of California Press: Oakland, CA, USA, 1976.
2. Ferrari, C.; Adamo, G.E. Autenticità e risorse locali come attrattive turistiche: Il caso della Calabria. *Sinergie* **2012**, *66*, 79–113.
3. Stamboulis, Y.; Skayannis, P. Innovation strategies and technology for experience-based tourism. *Tour. Manag.* **2003**, *24*, 35–43.
4. Turco, A. *Turismo e Territorialità. Modelli di analisi, strategie comunicative, politiche pubbliche*; Unicopli: Milano, Italy, 2012.
5. Hall, D.R.; Roberts, L.; Mitchell, M. *New Directions in Rural Tourism*; Ashgate: Hants, UK, 2005.
6. Long, P.; Lane, B. Rural tourism development. In *Trends in Outdoor Recreation, Leisure and Tourism*; Gartner, W.C., Lime, D.W., Eds.; CAB International: Wallingford, UK, 2000; pp. 299–308.
7. European Commission. Community action in the field of tourism. Commission communication to the Council. Bulletin of the European Communities, Supplement 4/86, p. 10. Available online: http://aei.pitt.edu/5410/ (accessed on 21 February 2015).
8. This concept resumes the sustainable development process, in which current needs are satisfied without compromising the needs of future generations (Brundtland Commission. *World Commission on Environment and Development Sustainability*; Oxford University Press: Oxford, UK, 1987).

9. Simonicca, A. *Antropologia del turismo, strategie di ricerca e contesti etnografici*; La Nuova Italia Scientifica: Roma, Italy, 1997; pp. 169–173.

10. Daugstad, K. Negotiating landscape in rural tourism. *Ann. Tourism Res.* **2008**, *35*, 402–426.

11. Barke, M. Rural tourism in Spain. *Int. J. Tour. Res.* **2004**, *6*, 137–149.

12. Lefebvre, H. The right to the city. In *Writings on Cities*; Kofman, E., Lebas, E., Eds.; Blackwell: Cambridge, MA, USA, 1995; pp. 63–184.

13. Corner, S.; Swarbrooke, J. *International Cases in Tourism Management*; Butterworth-Heinemann: Oxford, UK, 2004.

14. Cawley, M.; Gillmor, A.D. Integrated rural tourism: Concept and practice. *Ann. Tour. Res.* **2008**, *35*, 316–337.

15. Schubert, D. Active regions–shaping rural futures: A model for new rural development in Germany. In *Coherence of Agricultural and Rural Development Policies*; Diakosavvas, D., Ed.; OECD Publishing: Paris, France, 2006; pp. 333–352.

16. Zhou, L. Online rural destination images: Tourism and rurality. *J. Destin. Mark. Manag.* **2014**, *3*, 227–240.

17. Balestrieri, A.D. *Il turismo rurale nello sviluppo territoriale integrato della Toscana*; Irpet: Firenze, Italy, 2005.

18. Giffinger, R.; Fertner, C.; Kramar, H.; Kalasek, R.; Pichler-Milanović, N.; Meijers, E. *Smart Cities: Ranking of European Medium-Sized Cities*; Vienna University of Technology: Vienna, Austria, 2007.

19. Garau, C. Smart paths for advanced management of cultural heritage. *Reg. Stud. Reg. Sci.* **2014**, *1*, 286–293.

20. Turner, A. *Introduction to Neogeography*; O'Reilly: Sebastopol, CA, USA, 2006.

21. Goodchild, M.F. Twenty years of progress: GIScience in 2010. *J. Spat. Inf. Sci.* **2010**, *1*, 3–20.

22. Su, K.; Li, J.; Fu, H. Smart city and the applications. In Proceedings of 2011 International Conference on Electronics, Communications and Control, Zhejiang, China, 9–11 September 2011; pp. 1028–1031.

23. Basile, F. Una breve analisi congiunturale del Mezzogiorno nel contesto internazionale e riflessioni preliminari sul settore turistico. In *Turismo e territorio: l'impatto economico e territoriale del turismo in Campania*; Bencardino, F., Ed.; Franco Angeli: Milano, Italy, 2010; pp. 15–36.

24. Piano strategico per lo sviluppo del turismo in Italia. p. 10. Available online: http://www.agenziademanio.it/export/download/demanio/agenzia/5_Piano_strategico_del_Turismo_2020.pdf (accessed on 21 February 2015).

25. Angeloni, S. Cultural tourism and well-being of the local population in Italy. *Theor. Empir. Res. Urban Manag.* **2013**, *8*, 17–31.

26. Musarò, P. Responsible tourism as an agent of sustainable and socially-conscious development: Reflections from the Italian case. *Recerca: revista de pensament i anàlisi* **2014**, *15*, 93–107.

27. Sciarelli, S.; Della Corte, V. Il comportamento del turista in condizioni di forte incertezza decisionale. *Sinergie rivista di studi e ricerche* **2011**, *66*, 137–152.

28. Paniccia, P.; Vannini, I. Da impresa agricola a agriturismo: un percorso nell'ottica della sostenibilità. In *Le imprese nel rilancio competitivo del Made e Service in Italy: settori a confront*; Ciappei, C., Padroni, G., Eds.; Franco Angeli: Milano, Italy, 2012; pp. 59–71.

29. Piano strategico per lo sviluppo del turismo in Italia. pp. 5, 30, 47. Available online: http://www.agenziademanio.it/export/download/demanio/agenzia/5_Piano_strategico_del_Turismo_2020.pdf (accessed on 21 February 2015).

30. Santamato, V.R. Turismo e territorio: un rapporto complesso. In *Esperienze e casi di turismo sostenibile*; Messina, S., Santamato, V.R., Eds.; Franco Angeli: Milano, Italy, 2012.

31. Istat. Le aziende agrituristiche in Italia 2013. Rapporto aggiornato al 10 ottobre 2014. Available online: http://www.istat.it/it/archivio/133966 (accessed on 21 February 2015).

32. The ISTAT surveys consider only those farms that are treated as such, according to different regional standards. This means that these surveys do not consider those structures that provide the same services but are not incorporated because they are not considered as such by the regional laws.

33. Papotti, D. Marketing territoriale e marketing turistico per la promozione dell'immagine dei luoghi. *Rivista Geografica Italiana* **2006**, *113*, 285–306.

34. Go, F.; Della Lucia, M.; Trunfio, M.; Martini, U. Governing sustainable tourism: European networked rural villages. In *Rural Cooperation in Europe: In Search of the "Relational Rurals"*; Kasabov, E., Ed.; Palgrave Macmillan: Chennai, India, 2014; p. 163.

35. Puggioni, G.; Atzeni, F. I comuni sardi a rischio di estinzione. In *Comuni in estinzione. Gli scenari dello spopolamento in Sardegna*; Progetto IDMS—2013; Regione Autonoma Della Sardegna: Cagliari, Italy, 2013; pp. 15–45.

36. Elaborations made by the author on Demoistat Data (2014).

37. Barumini, Collinas, Furtei, Genuri, Gesturi, Las Plassas, Lunamatrona, Pauli Arbarei, Sanluri, Segariu, Setzu, Siddi, Tuili, Turri, Ussaramanna, Villamar, Villanovaforru, and Villanovafranca.

38. Bonifazi, C.; Heins, F. Le dinamiche dei processi di urbanizzazione in Italia e il dualismo Nord-Sud: un'analisi di lungo periodo. *Rivista economica del Mezzogiorno* **2001**, *15*, 713–748.

39. Calza, B.P.; Cortese, C.; Violante, A. Interconnessioni tra sviluppo economico e demografico nel declino urbano: il caso di Genova. *Argomenti* **2010**, *29*, 105–131.

40. Dematteis, G. Montagna e aree interne nelle politiche di coesione territoriale italiane ed europee. *Territorio* **2013**, *66*, 7–15.

41. Author's elaborations on census data from 1861 to 2011. In *La popolazione dei comuni sardi dal 1688 al 1991*; Angioni, D., Loi, S., Puggioni, G., Eds.; Cuec: Cagliari, Italy, 1997.

42. Author's elaborations on Data DEMO-ISTAT 2014 and on Urbistat's Data. Available online: http://www.urbistat.it (accessed on 21 November 2014).

43. Author's elaborations on Urbistat's Data. Available online: http://www.urbistat.it (accessed on 5 March 2015).

44. For example, the Touristic Consortium of *Genna Maria* in the municipality of Villanovaforru was the first such association. This story begins in 1969, but was only legally recognized in 1982 because of the lack of agreement among the different municipalities. The Consortium, that later changed its name to *Sa*

Corona Arrubia, then had members from other municipalities. Today there are about twenty members, and almost the same number of museums, as well as thirty-two archaeological areas, and a park's environmental interest. Today *Sa Corona Arrubia* is having difficulties, although numerous other Consortia (the Consortium of Two Jars; the Consortium of the Natural Park of Monte Arci; and the Consortium *Sa Perda e'Iddocca*) have joined it over time. All these Consortia are directed and oriented by , and the interprovincial Local Action Groups (LAG) of the Union of Communities of Marmilla, development agencies; the interprovincial Local Action Groups (LAG) of Marmilla Borgioli C., Deligia M.G., Sa Corona Arrùbia—Consorzio Turistico della Marmilla. Available online: http://sistemimuseali.sns.it/content.php?idSC=103&el=9&c=106&ids=3&o=sistemiCulturali_dataInizioInterna#ds1 (accessed on 11 May 2015).

45. Sirena, P. Museo Sa Corona Arrubia Museo Naturalistico del Territorio G. Pusceddu. Available online: http://www.sacoronarrubia.it / (accessed on 4 February 2015).

46. In 2005 the World Tourism Organization (WTO), used the "three pillars" to describe sustainable tourism: "sustainability principles refer to the environmental, economic and socio-cultural aspects of tourism development, and a suitable balance must be established between these three dimensions to guarantee its long-term sustainability". *Making Tourism More Sustainable: A Guide for Policy Makers*; United Nations World Tourism Organization (UNWTO) and the United Nations Environment Programme (UNEP): Madrid, Spain, 2005; pp. 11–12.

47. Fons, M.V.S.; Fierro, J.A.M.; y Patiño, M.M. Rural tourism: A sustainable alternative. *Appl. Energ.* **2011**, *88*, 551–557.

48. Rural Development Programme (RDP) 2014–2020. Available online: http://www.regione.sardegna.it/speciali/programmasviluppororale/2014-2020/psr-2014-2020 (accessed on 8 May 2015).

49. Regional Development Plan (RDP) 2014–2019. Available online: http://www.regione.sardegna.it/j/v/66?s=1&v=9&c=27&c1=1207&id=44429 (accessed on 8 May 2015).

50. Belij, M.; Veljković, J.; Pavlović, S. Role of local community in tourism development: Case study village Zabrega. *Glasnik Srpskog geografskog drustva* **2014**, *94*, 1–14.

51. Hall, M.C.; Allan, W. *Tourism and Innovation*; Routledge: London, UK, 2008.

52. "The structural dimension (network ties, configuration and stability) reveals non-hierarchical and dense ties [...]. Community members' relationships are direct, informal, and long-term. There social ties feed and consolidate a strong sense of belonging to the place where these communities live and work and are the base on which inter-member economic ties and knowledge sharing are developed". Go, F.; Della Lucia, M.; Trunfio, M.; Martini, U. Governing Sustainable Tourism: European Networked Rural Villages. In *Rural Cooperation in Europe: In Search of the 'Relational Rurals'*; Kasabov, E., Ed.; Palgrave Macmillan: Basingstoke, UK, 2014; pp. 173–174.

53. "From a cognitive dimensional perspective, social capital suffers from the fragmentation inherent in the heterogeneity of local stakeholders and sectorial diversification. Diversity of aims, interests, and competence renders the establishment of a critical mass for group decision-making among the local stakeholders difficult. [...] Information and education could help increase the awareness that networked knowledge sharing facilities inclusive economic institution-building and the emergence of a virtuous cycle of value-adding processes". Go, F.; Della Lucia, M.; Trunfio, M.; Martini, U. Governing Sustainable Tourism: European Networked Rural Villages. In *Rural Cooperation in Europe: In Search of the 'Relational Rurals'*; Kasabov, E., Ed.; Palgrave Macmillan: Basingstoke, UK, 2014; p. 175.

54. "The relational dimension of social capital focuses on the character of connections, which serve to reinforce not only an organisation's internal logic, but also (external) trustworthy relations". Go, F.; Della Lucia, M.; Trunfio, M.; Martini, U. Governing Sustainable Tourism: European Networked Rural Villages. In *Rural Cooperation in Europe: In Search of the 'Relational Rurals'*; Kasabov, E., Ed.; Palgrave Macmillan: Basingstoke, UK, 2014; p. 176.

55. Nasser, N. Planning for urban heritage places: Reconciling conservation, tourism, and sustainable development. *J. Plan. Lit.* **2003**, *17*, 467–479.

56. Weaver, D.B. Tourism and the elusive paradigm of sustainable development. In *A Companion to Tourism*; Lew, A.A., Hall, C.M., Williams, A.M., Eds.; Backwell: Oxford, UK, 2004; pp. 510–524.

57. Law, R.; Leung, R.; Buhalis, D. Information technology applications in hospitality and tourism: A review of publications from 2005 to 2007. *J. Travel Tour. Market.* **2009**, *26*, 599–623.

58. Garau, C.; Ilardi, E. The "Non-Places" Meet the "Places:" Virtual Tours on Smartphones for the Enhancement of Cultural Heritage. *J. Urban Tech.* **2014**, *21*, 79–91.

59. Garau, C. From Territory to Smartphone: Smart Fruition of Cultural Heritage for Dynamic Tourism Development. *Plan. Pract. Res.* **2014**, *29*, 238–255.

60. According to Porrello, cultural planning enhances the totality of cultural, environmental, and naturalistic goods by taking strategic actions that are in line with the economic and social sustainability of the local context under study. Porrello, A. *L'arte Difficile del Cultural Planning*; Grafiche Veneziane: Venice, Italy, 2006; Available online: http://www.iuav.it/Ateneo1/docenti/pianificaz/docenti-st/Antonino-P/materiali-/Cultural_Planning.pdf (accessed on 11 March 2015).

61. Below are some examples. Available online: http://www.turismo.intoscana.it/site/it/itinerario/Sei-un-turista-rurale/ and http://www.iAGRITURISMO.it. This last link in particular allows users to view the portal of Italian Farms in i-mode, with any type of mobile phone. Here too, the Tuscany region ranks high in the number of farmhouses present.

62. Mistretta, P. Beni Culturali e sistema territorio. *ArcheoArte* **2012**, *1*, 11–19.

63. Mureddu, D.; Murru, G. *Alla scoperta dei monumenti della Marmilla: Archeologia, arte, architettura*; CRES: Cagliari, Italy, 2000.

64. Lilliu, G. *La civiltà dei Sardi dal paleolitico all'età dei nuraghi, (Vol. 98)*; Nuova Eri: Torino, Italy, 1988.

65. Atzeni, C., Ed.; *I manuali delle colline e degli altipiani centro-meridionali*; Tipografia del Genio Civile: Roma, Italy, 2009.

66. Atzeni, C., Sanna, A., Eds.; *Architettura in terra cruda*; Tipografia del Genio Civile: Roma, Italy, 2009.

67. Purpura, A.; Ruggieri, G. Il distretto turistico: Caratteristiche e modelli organizzativi. In *I distretti turistici: Strumento di sviluppo dei territori. L'esperienza nella regione Sicilia Milano*; Cusimano, G., Parroco, A.M., Purpura, A., Eds.; Franco Angeli: Milan, Italy, 2014; pp. 19–41.

68. McAreavey, R.; McDonagh, J. Sustainable rural tourism: Lessons for rural development.*Sociologia ruralis* **2011**, *51*, 175–194.

10
CHAPTER

Critical Sustainability: Setting the Limits to Growth and Responsibility in Tourism

Jarkko Saarinen [1,2]

[1] *Department of Geography, P.O. Box 3000, FI-90014 University of Oulu, Finland; Tel.: +358-404-840-777*

[2] *School of Tourism and Hospitality, University of Johannesburg, P.O. Box 524, Auckland Park, Johannesburg, South Africa*

ABSTRACT

The idea of sustainable development has been discussed in tourism research for almost a quarter of a century. During that time, sustainability has become an important policy framework for tourism and regional developers guiding their planning and development thinking. Sustainability has also emerged academically as an important field of research with an emphasis on defining the limits to growth and responsibilities in tourism. However, while there are urgent needs to incorporate sustainability into tourism, there is also a growing amount of frustration among scholars on the conceptual nature of sustainability and how tourism as a private-driven economic activity relates to the ideals of sustainable development. This has created an increasing need to understand and potentially reframe the concept. The purpose of this paper is to overview the conceptual dimensions of sustainable tourism and discuss some of the main sources of frustration. Based on this, it is concluded that while a conceptual plurality seems to be unavoidable, there is a need to re-frame *i.e.*, rescale and decentralize tourism in policy frameworks and practices aiming towards sustainability.

Keywords: sustainable tourism; responsible tourism; limits to growth; governance

1. INTRODUCTION

Sustainability has emerged as a paradigm in tourism planning and development. The demand for sustainability in tourism is based on several interrelated processes. Since the 1960s and 1970s, the growth of global tourism has been intensive, with significant impacts on the tourism system in a global-local nexus. These impacts and especially the negatively perceived outcomes of tourism growth in destination regions highlighted a need to guide and limit tourism growth in many places [1,2,3]. In addition, the societal context in the late 1960s and 1970s supported the focus on environmental concerns and global inequalities: the emergence of the environmental movement, the limits to the growth debate (see [4]) and a north-south divide in global politics, for example, had a strong influence on the discussion on the nature of tourism development [5]. Rather than focusing on the limits to growth or inequalities in global tourism, however, the discussions were more grounded on a destination or site scale analysis (see [6]).

The political conceptualisation of sustainable development presented in the Brundtland Commission's report "Our Common Future" in 1987 [7] further fuelled the policy aims and needs to limit the impact of tourism. By the early 1990s, the idea of sustainability was built into tourism development [8,9] with an evolving number of conceptualisations and a search for indicators (see [10,11,12,13]). At the same time, the transformation of production and consumption in Western societies towards post-Fordist production supported alternative and 'green' trends in consumption, including the tourism sector. This resulted in a number of new forms and terms such as ecotourism and responsible tourism [14,15] which were often seen as more 'environmentally conscious' products [16,17]. Recently, sustainability in tourism has been linked to ethical consumption [18,19], including high level policy aims to reduce global poverty, for example (see [20,21]). Related to this, the Secretary-General of the UNWTO, for example, has stated that the global tourism industry could play a major role in the achievement of the United Nations Millennium Development Goals (UN MDGs) by 2015 [22]. Similarly, the World Bank [23] has emphasised the transformative role that tourism might play in developing economies and poverty alleviation and/or elimination.

Indeed, the tourism industry has become responsible for various aims and impacts on global and local scales. In general, responsible tourism refers to tourism development principles and practices aiming to make places better for people to live and visit [24]. It aims to minimise the negative and maximise the positive social, economic and environmental impacts of tourism in destination communities and environments by promoting ethical consumption and production among all stakeholders. Although the concept of responsible tourism is often used as a specific form of tourism, its principles and guidelines are rather similar to the general aims of sustainable tourism: according to Richard Sharpley ([25], p. 385) "it is difficult, or even impossible, to distinguish responsible tourism from the concept of sustainable tourism". He has also critiqued whose responsibility we are referring to when using the concept. Indeed, it is important to identify who actually are responsible and for what, *i.e.*, are we, the industry, governmental agencies *etc.*, responsible to local communities, societies and people in general, for example, when aiming to develop and manage tourism in a responsible way.

While responsibility is partly built on the same grounds as sustainability in tourism with over-lapping targets, there is a contextual difference. Responsibility discourse is also a product of neoliberal "self-organising" modes of new governance with resulting corporate social responsibility (CSR) initiatives and the creation of a "perfect green consumer" who does not consume less but consumes in a responsible way (see [16,26]). An increased focus on responsibility also has an academic grounding. In geography, for example, Smith [27] has discussed "the moral turn" in human geography and Lawson [28,29] and Gregory [30] have emphasised a need for caring not only for "our own" but for distant strangers, which resonates well with the recent discussions on responsible tourism and especially volunteer tourism (see [31]), for example, aiming to social development goals in destination communities.

Overall, the need for sustainability in tourism is an outcome of a complex set of processes, systems, aims and scales (see [32]). This has led to a multifaceted conceptual ground of sustainable tourism and a criticism concerning the clarity and meaning of the concept and its applicability in tourism (see [33,34,35,36]). Swarbrooke ([37], p. 41) has stated that "sustainable tourism is perhaps an impossible dream" and some scholars have even suggested that we should actually get rid of the concept and idea of sustainability in tourism. According to Sharpley ([38], p. xiii), for example, it is time to move beyond sustainability as the "academic study of sustainable tourism development has reached something of an impasse". While criticisms and evolved frustrations are understandable, the demand for sustainability, responsibility and limits to growth thinking in tourism has not disappeared; in fact, the reality is quite the opposite. According to the World Bank ([23], p. 11) "tourism is a three-billion-a-day business", the growth of which is estimated to reach 1.6

billion international tourists arrivals by 2020 (UNWTO, 2013). As further noted by the World Bank ([23], p. 7) the tourism industry and its growth "comes with its own set of risks and challenges". Therefore, the need for the idea of sustainability in tourism is now more urgent than ever before.

Instead of actively sinking the idea of sustainability, there is a need to further reflect on the conceptual idea of sustainability, including the issues of responsibility, in tourism and evaluate the potential reasons why we have failed to progress faster in sustainable tourism research and operations in practice over the past 25 years. The purpose of this paper is to critically overview the conceptual dimensions of sustainability in tourism based on the literature and previous discussions. As the focus is on sustainability on tourism, the reviewed literature is mainly based on (sustainable) tourism studies with links to general sustainable development and governance literature. After a short conceptual overview, some of the main sources of criticisms and frustrations are discussed. Finally, the critical points of departure in the sustainability in tourism and development are evaluated.

2. SUSTAINABLE TOURISM

2.1. Sustainable Development in Tourism

The general idea of sustainable development has been described as complex, vague and difficult to operationalize in practice, and these elements have followed many of the later definitions of the World Commission on Environment and Development's (WCED) [7] original formulation (see [39,40]). While the concept involves analytical weaknesses, it has also provided a common platform on which different stakeholders in development can interact, negotiate and reflect on the consequences and limits of their actions for the social, economic and ecological environment [6]. Due to conceptual ambiguity, though, many scholars in tourism research have stated that there are no exact working definitions of sustainable tourism, which has partially made progress in research slow (see [2,11,41]).

The United Nations Environment Programme (UNEP) and the World Tourism Organization (WTO) ([42], p. 12) have defined sustainable tourism as "tourism that takes full account of its current and future economic, social and environmental impacts, addressing the needs of visitors, the industry, the environment and host communities". Similarly as the "mother-concept" of sustainable development, the above definition is value-laden and open to multiple interpretations and perspectives with corresponding references for the rhetoric of sustainable development as a holistic, future-oriented and socially equal global scale process [32,38,43]. All this has resulted in different understandings of, and perspectives on, sustainability in tourism and how the limits to growth are defined in tourism. Next, some of the main perspectives of setting the limits to growth in tourism are discussed based on the previous categorisation (see [6]).

2.2. Different Traditions of Sustainable Tourism

Based on the existing studies on sustainable tourism, several interrelated traditions focusing on the limits to growth in tourism can be identified. These traditions, which have been previously termed resource-based, activity-based and community-based sustainability in tourism (see [6]), differ in focus and in their understanding of, and relation to, resources used in tourism (Table 1). The traditions are characterized by different ideas on the nature of the limits to growth and different perspectives on these limits and how they can be known and defined. Although the impacts of tourism are increasingly global and operating in the whole tourism system (involving destinations, the generating regions of tourists and connecting routes), the main emphasis of past and existing research, management and policy activities has been largely based on the destination (*i.e.*, local) level.

Table 1. Traditions of sustainable tourism and setting the limits to growth.

	Resource-Based	Activity-Based	Community-Based
Origin/Manifestation	Carrying Capacity Model	Product Cycle (TALC)	Participatory Planning
Orientation	Environment (physical)	Industry	Community
Limits to growth	Objective/Measurable	Relative/Changing	Constructed/Negotiated
Resource and system view	Static	Dynamic	Dynamic/Static
Time scale	Long	"Now"	Short

The initial conceptualizations of the limits to growth in tourism were related to the carrying capacity model. It has its origins in the 1920s and 1930s' tourism and recreation studies (see [44,45]) but the major research boom materialized in the 1960s and 1970s and continued until the early 1980s (see [46,47,48,49,50]). While carrying capacity thinking may be regarded as a historical phase in tourism studies, it is still occasionally interpreted as an application of sustainable tourism (see [34,51]) and, thus, the concept occupies a key position in understanding the limits to growth thinking in tourism [52,53]. In research, the carrying capacity model led to a search for the 'magical' number of tourists in a certain space (and time period) which cannot be overstepped without causing serious negative impacts on the resources [54,55,56]. This resource-based tradition has its roots in the natural sciences and positivism, implying an objective and measurable limit or stage of growth at which there is no room for any more tourists or tourist activities in a certain environment [57]. In order to achieve further growth, tourism actors will have to cope with the environment in a better way without significantly changing the resource and its integrity. Thus, the limits to growth and impacts of tourism are evaluated in relation to the resources used and the assumed or known natural or original (non-tourism) conditions (see [58,59,60]). Obviously, the challenges are how to define the original non-tourism conditions of the resources or how to separate the impacts of tourism from changes caused by other activities and natural or human-induced processes taking place the same space (see [61]).

Tourism development always causes some impacts, which leads to the critical question of which impacts are 'objectively' acceptable and to what degree [6]. Indeed, tourism is a dynamic activity that transforms its resources and changes their capacity to absorb tourism via management actions and product development (see [32,54]). Therefore, the resource-based focus as a 'static' limit has turned out to be problematic for the industry. In this respect, McKercher [62] has emphasized that the tourism industry actually needs to play an active role in defining sustainability in tourism (see also [63]). In this respect, the industry and related institutions, such as the World Tourism Organization, have been actively involved in defining the goals and principles of sustainability and related policies (see [9,42]). These industry perspectives refer to an activity-based tradition of sustainable tourism. In contrast with resource-based sustainability, tourism actors do not primarily alter their behaviour based on the static interpretation of the limits to growth: in order to grow, the industry and other related actors can, and quite often do, modify the environment for their economic development purposes.

The activity-based tradition is industry-oriented. It refers to tourism-centric approaches in tourism development discussions [64], focusing on the needs of tourism as an economic activity [38] and its resource base [37]. From this perspective, the limits to growth are not primarily based on the capacity of the destination and its original resources for absorbing tourism, but on the industry and its capacity or incapacity to generate growth. This approach is illustrated in the evolution of the tourism area life cycle model [65], in which the relationship between life cycle—representing tourism growth—and the limits of carrying capacity is a dynamic one [66,67]: after a stagnation phase, indicating that the limits of carrying capacity have been reached, the development of a tourist destination may actually be activated again based on new products, infrastructure development and marketing, for example. Thus, the activity-based tradition is grounded on a relational approach and understanding of space (destination), which implies that certain tourism activities, tourist segments or products may have different kinds of limits to their growth and their ability to absorb increasing numbers of tourists. A non-growth situation implies that the limits to growth are reached and modifications are needed in tourism products in order to achieve further growth (see [65,68]).

In this respect, the relation between resource-based and activity-based sustainability can be quite conflicting. As the number of tourists increases and the destination evolves (*i.e.*, grows) cyclically through changing modifications of destination as a product, indicating that the limits to activity-based sustainability have not yet been reached, tourism growth may overstep some of the resource-based limits to change [6]. In order to overcome the potential and often highly probable conflicts between the industry, other stakeholders and resource use, various participation processes and governance models have been used and developed. These processes refer broadly to community approaches in tourism studies [69,70] in which the setting of limits to growth is based on participation and negotiations (see [71,72]). This community-based tradition aims to involve communities and other (local) stakeholders in tourism development and management by stating that communities should have control over the uses and benefits of (common) resources used in tourism (see [73,74,75]). Thus, in order to reduce the negative impacts of tourism and safeguard effective benefit sharing, local participation, awareness creation and control over tourism development are said to be needed [76,77,78].

The community-based tradition strongly implies that the limits to growth are socially constructed (see [79,80,81]). This does not indicate that the limits are fluid or open to any given definition but they are identified by a wider set of actors than the industry or the environmental issues alone, for example. From the community-based perspective, sustainability refers to the maximum levels of the known (or perceived) impacts of tourism that are permissible in a certain time-space context before the negative impacts are considered to be too disturbing from the perspectives of specific social, cultural, political or economic actors who possess sufficient knowledge and power over the chosen indicators and criteria [6]. Thus, the issues of power and knowledge are in a key position: while tourism impacts do exists in a physical "reality" (e.g., in resource-based world) outside human values and perceptions, in the world of meanings and human preferences, the question of whether these changes are acceptable or unacceptable depends on specific (societal and/or individual) values, attitudes, knowledge and priorities concerning the role and impacts of tourism (see [32,82,83]). All these are linked to the idea of power, *i.e.*, an "ability to impose one's will or advance one's own interests" in development as stated by Reed [84]. According to him, power comes across as an instrument to be managed and based on this, it is necessary to safeguard an equitable power sharing among different stakeholders in tourism. The community-based tradition aims to empower specifically the host communities in tourism development and the governance of key resources [85]. Thus, for community-based sustainability, the determination of the limits to growth is associated with power relations constituted by different stakeholders, which are constantly transforming, making the decision-making process challenging [86].

3. SOURCES OF FRUSTRATIONS IN SUSTAINABLE TOURISM

Since the 1990s, the idea of sustainable tourism has charmed academics, tourism developers, NGOs and policy-makers. However, almost from the beginning of sustainable tourism discussions it has also attracted harsh criticism [34,62] and recently, there have been increased calls for going beyond the sustainable development paradigm in tourism studies (see [38]). Indeed, there are many problems involved in sustainable tourism and especially in relation to sustainable development—issues which have been widely discussed elsewhere (see [6,33,35,36,87]). Instead of repeating the existing critical debates in full, as they are generally known and discussed in tourism studies, the following key sources of frustration are briefly discussed and re-framed: conceptual vagueness, ideals *versus* reality in practice, and locating responsibility.

Similar to the idea of sustainable development, the concept of sustainable tourism is said to be vague [34,35,36,43]. Indeed, several conceptualisations exist and some of them are potentially conflicting with each other and the discussed traditions of setting limits to growth in tourism. Swarbrooke ([37], p. 13) for example, has defined sustainable tourism as "tourism which is economically viable but does not destroy the resources on which the future of tourism will depend, notably the physical environment and the social fabric of the host community". While it is a solid definition, it evidently focuses on the needs of the tourism industry and the sustainable use of its resources (see [88]), *i.e.*, representing the activity-based tradition of sustainability aiming to sustain tourism. Burns [64] has termed this kind of view of tourism development as the "Tourism First" approach, which indicates that developing the industry is the focus of evaluations and planning.

Various perspectives and competing ideas of the concept of sustainable tourism can be seen as problematic. However, while they are challenging for researchers and policy-makers, it is good to note that there are very few, if any, conceptual definitions in the social sciences we universally accept as being "the definition" and the only existing way to understand an issue. In addition, a value-based nature of sustainable tourism has been regarded as a major challenge for the concept (see [36,38]). Indeed, sustainable tourism is value-laden idea, but in contrast to physics and other natural sciences, for example, concepts in the social sciences in general tend to be loaded with values. They are also used in academic and societal discussions and politics based on different kind of ideologies and socio-economic and cultural preferences and aims. By replacing sustainability with some other term pre-perceived as less value-based may not change the situation: as soon as adaptation or resilience, for example, were applied from natural science contexts to the social sciences, their definitions have multiplied and become transparently value-laden and contested (see [89,90,91]).

Thus, instead of criticizing the conceptual plurality or aiming for one universal value-free definition of sustainable tourism, it would probably be more fruitful and realistic to aim to recognize what kind of basic assumptions and preferences there are behind the utilization of the concept and setting the limits to the growth in tourism development in specific situations. However, this does not remove the other key problem of the conceptualization(s) of sustainable tourism: while it rhetorically refers to sustainable development, it seems to fail to deliver a similar kind of holistic framework [36], especially in relation to spatial and temporal scales. While tourism is a global scale economy and a system operating in "home regions" and "destinations" that are linked together by the means of routes and transit regions, the focus of sustainability in tourism has mainly been on tourist destinations or certain sub-parts and sites of destinations. As stated by Inskeep ([92], p. xviii), for example, "the sustainable development approach can be applied to any scale of tourism development from larger resorts to limited size special interest tourism..." This destination and certain tourist activity emphasis in sustainable tourism may grasp the most visible processes and impacts related to tourism, but only a fragment of the total (see [6,93]). In the era of climate change concern, for example, the destination view is limited and seriously problematic. Gössling *et al.* ([94], p. 208) have stated in their ecological footprint analysis of tourism in the Seychelles that "efforts to make destinations more sustainable through the installation of energy-saving devices or the use of renewable energy sources can only contribute to marginal savings" (to the environment) as 97% of the footprint is a result of air travel!

In addition to being a kind of a red herring with misleading clues to global scale responsibility, the ideals of sustainable tourism are difficult to put into place in practice [36]. These ideals involve, for example, the same set of three elements and principles as sustainable development: the elements of ecological, social and economic sustainability and the principles of holism, futurity and equity. However, tourism is often seen as an industry that principally places more emphasis on the economic dimension and the present context [64,95]. The economic preference and the predestined focus in the industry is well implied in Getz's ([96], p. 24) definition of a touristic space as being "an area dominated by tourist activities or one that is organized for meeting the needs of visitors". This means that the needs and values of the customers (non-local people) and the industry are the leading guidelines in market-driven economic activities like tourism. While the tourism-centric view is understandable, as without economically viable businesses there is no tourism (including sustainable tourism), it basically challenges the ideal of the equity between the elements and principles of sustainability in tourism. As noted by Eagles ([97], p. 231), "financial efficiency may be a pivotal criterion given more importance" in the governance of tourism and recreation in protected areas, such as national parks. Thus, in tourism the three "pillars" of sustainability are not often equal. Interestingly, in the context of recent discussions on resilience, which is occasionally interpreted as being the "new sustainability", a similar preference for economic issues has been noted. A United Nations panel Resilient—Resilient Planet ([98], p. 12) indicated that the reason why the ideals of sustainability have not been implemented in practice is that "sustainable development has not yet been incorporated into the mainstream national and international economic debate". What the panel indicates is that the economic pillar (element) is often "higher" than ecological or social ones, *i.e.*, economic issues are more relevant "now" when development needs and decisions are made. This problem and focus on the current (and local scale) context in policy-making and development practices in sustainable tourism, sustainable development or environmental politics, in general, can be seen as a "sliding present trap". What the term means is that while economic issues are prioritized in the present day context, the environmental and wider elements of sustainability are seen to become more and arguably equally important in the future as the economic aspects in development. However, this equal future is like a horizon keeping its distance from the sliding present day context where decisions are (continuously) made and justified.

The sliding present trap makes the challenges to implementing the known or estimated future costs of growth to present day decisions and practices understandable (e.g., in tourism-climate change relations) (see [99]). This also relates to another source of frustration in sustainable tourism research and debates: responsibility and the question of whose responsibility sustainability in tourism development is [25]. There are basically two interrelated views on the issue of responsibility. On one hand, consumers are seen to lead the industry towards more sustainable operations due to increasing environmental awareness in societies, which is assumed to evolve towards higher demand for sustainable products. Changing modes of consumption referring to new tourism and higher environmental responsibility of the new tourists (see [17]) are further supporting the assumption. On the other hand, the industry is said to be adopting sustainable development principles and is therefore seen to progress towards responsible modes of production in tourism. This line of thinking is manifested in the Tourist Operators Initiative to Sustainable Tourism [100], for example, which set self-organized rules and indicators for tourism businesses. In addition, as indicated in the above, the assumed higher environmental awareness of the customers and estimated increasing demand for environmentally friendly products are seen to trigger the supply of sustainable tourism products (see [101]).

The logic of progressive responsibility in tourism consumption and production is rather coherent. However, previous studies indicate challenges in both views of responsibility. According to Sharpley [17], most academic studies show that despite "greening" attitudes, environmental concerns still remain low in tourists' consumption priorities (see also [102,103,104]). In terms of tourism businesses adopting sustainable tourism principles, there has been improvement. However, the adopted or modified principles are often said to be "flexible" *i.e.*, industry-oriented, and the evaluation criteria used do not deeply reflect the ideals of sustainability (see [32,34,64,105]). The sustainable tourism strategy of Levi resort in Finnish Lapland, for example, states that the goal for social sustainability is that 75% of the employees working in the resort are local people. Compared to many other destinations in Finnish Lapland, however, it is not a remarkably "good" number [106]. Thus, the question is: does it represent sustainability or is the limit set to a level that perhaps more or less legitimizes the current situation at the destination [107]? In addition to the problems with self-organized limits to growth and responsibility, only a relatively small number of businesses are actually aiming high or even beyond their legal obligations [25].

4. CONCLUSIONS

4.1. Discussion

The idea of sustainability is important for the tourism of today and especially for its future. However, the criticism of the conceptualization and operationalization of sustainable tourism has recently increased. Still, deepening frustrations have yet to create viable alternatives to sustainability as a paradigm, and sustainable development and sustainable tourism are still in a high profile position in international and national politics, development and planning debates, and also in research. Nevertheless, the expressed criticism has relevant grounding, which underlines a need to understand and potentially re-frame the concept of how the limits to growth are set in tourism and how to overcome the perceived shortcomings of sustainable tourism in policy and practice.

A conceptual plurality seems unavoidable. Like most of the conceptual agreements in the social sciences, the meaning of sustainable tourism as a value-laden idea is—at least partly—dependent on the development paradigm and use context (see [108]). The overviewed three traditions have their origins in tourism and environment discussions, and also outside sustainable development thinking. The resource-based tradition reflects the limits of the original conditions of the resources used in tourism and the needs to protect the natural and socio-cultural environment from unacceptable changes caused by tourism activities. Contrast that to the activity-based tradition, which refers to the resource needs of the industry with aims to sustain the economic capital invested in tourism development. The community-based tradition strives to find a balance between tourism industry and different stakeholders, especially local communities, by emphasizing local involvement, control and empowerment in tourism development. All these perspectives have their advantages in different use contexts and they can complement each other, but in respect to the idea of sustainability and the future challenges of humanity, they all share the same major limitation, which is the strong focus on the local scale.

Thus, while we may need to accept the conceptual plurality and recognize the importance of context in research, there is a need to re-frame sustainability in tourism in a more critical manner. The key question is on what conditions sustainable tourism could represent sustainable development beyond the local scale, *i.e.*, in a local-global nexus. In the future, the local solutions to global challenges currently being sought and developed are not enough. Transferring ideals into practice by considering that the limits to growth on a global scale is a challenging but much needed exercise with an obvious probable conclusion ahead, implying that tourism involving a major component of air travel, e.g., long-haul flights, cannot be sustainable [94]. This kind of conclusion is in conflict with the currently hegemonic idea of sustainable tourism, stating that all kinds of tourism can aim (realistically) to achieve sustainability in their operations (see [11]). The current implicit scaling of the limits to growth being potentially locally sustainable (e.g., ecotourism operations in developing countries far away from markets) but globally probably unsustainable is neither responsible nor sustainable. Sustainability in tourism is a matter of both local and global responsibilities.

In addition to the stronger recognition of sustainability on a global scale, there is a need to relocate tourism in sustainable tourism discussions. Instead of tourism-centric or "Tourism First" approaches, tourism as an economic activity needs to be decentralized *i.e.*, repositioned in the discourses and practices referring to sustainable development (see [15,33,109,110]). This means that instead of taking the (central) role of tourism as granted, the industry is rather seen as a potential tool for sustainable development—but not as representing sustainable use of resources *per se*. Thus, tourism is an option that could be used for sustainable development. The challenge of tourism-centric evaluations is not only an issue in sustainable tourism but more widely in tourism development-related discussions. Tourism's role in regional development, for example, is often seen as positive, and tourism is stated to be used as a vehicle for regional development in various policy-documents. In this context, regional development is usually evaluated in terms of tourism employment, tourism revenue and tourist flows, *etc*. However, regional development (compared to growth) involves deeper and qualitative goals, referring to an improvement in the quality of life and well-being of the people, (see [107,111,112]) which are not automatic results of tourism growth indicators. Thus, based on the original conceptualization of sustainable development [7], sustainability in tourism development should primarily be connected with the needs of people—not a certain industry—and the use of natural and cultural resources in a way that will safeguard human needs and provide quality of life and well-being in the future (see [40,111]). Therefore, in addition to the question "whose responsibility", the search for responsibility in tourism should ask "who are we responsible to" in sustainable tourism development.

Obviously, the needs of people and the tourism industry are not necessarily conflicting but as Ringer ([95], p. 9) has indicated, tourism is an industry "that satisfies the commercial imperatives of an international business, yet rarely addresses local development needs". Thus, as noted above, tourism-led economic growth does not necessarily translate into benefits for the people and their environment. Therefore, the order of preference between serving and sustaining resources for people and the tourism industry should be clear and in logical relation to the idea of sustainable development. Tourism can be a potential and fruitful tool for sustainable development but it may not always be the most favorable use of resources in specific locations (see [87,109,112]) and locally "sustainable tourism" may in practice be unsustainable globally and/or locally in the long term.

It is obvious that "the source of the rule" [113] in setting the code of conduct for, and limits to, sustainable tourism is mainly the industry itself. In that self-organizing context, however, the search for responsibility in tourism has demonstrated "a market failure", indicating than an external intervention may be urgently needed. According to many previous studies, the demand for environmentally-friendly products has not yet materialized substantially enough and the industry's progress in sustainable tourism beyond tourism-centric and economic criteria has been relatively slow. Thus, the ongoing new governance that is hollowing out state responsibility and emphasizing corporate, non-governmental or citizen (customer) responsibilities (see [114]) does not seem to be working (fast enough) for sustainable development in tourism. Therefore, in order to have tools for setting the limits to growth in tourism in a local-global nexus with less tourism-centric evaluation criteria, stronger governmental and inter-governmental policies and regulations are most probably needed (see [115]). The governance of sustainable tourism should also implement future costs (see [116,117]) to present day evaluations and see beyond short-term economic prospects. This is a challenging but unavoidable task if the tourism industry aims to be guided towards sustainability.

4.2. Concluding Remarks

To summarize, while there are major challenges in sustainable tourism as a concept and development tool in practice, the calls for forgetting the idea may be premature. Instead of going beyond sustainability in tourism, there may be a real need to take serious steps back towards the original ideas of sustainable development. Therefore, the re-framing of sustainable tourism as a less tourism-centric activity operating in a local-global nexus is vital. Sustainability is a crucial element for the future of tourism and this re-framing would make it a critical tool and dimension for the evaluations of the limits to growth in tourism with strong references to evolving discussions on ethical components in tourism (see [118,119,120]).

The ethical element in sustainable tourism development is built upon both theory and practice. This means that the industry would need to change and reconsider its position in development discourse if it really is aiming to promote sustainable development in tourism beyond rhetoric and "green washing". However, to expect the industry as a private sector economic actor to substantially share its benefits, and decentralize its own role and position in its own operations may not be realistic. As Scheyvens [121] has critically asked in the context of tourism and poverty alleviation: why should we assume that the tourism industry has some ethical commitment to ensuring that their operations contribute to the alleviation of poverty (see also [122])? While there are individual companies doing well in this respect, a large majority seem to continue to operate along the activity-based limits to growth. Therefore, the industry as a whole and its customers need to have firmer guiding regulative frameworks for creating a wider responsibility and a path towards sustainable development. Obviously, this is easier said than done, which is evident in the global scale regulative processes such as the Kyoto Protocol, for example. However, the first step is to recognize the need for re-framing sustainability in tourism: after that, the operationalization of firmer regulative frameworks is definitely easier to process than under the current hegemonic idea of sustainable tourism as a local-scale and self-organized industry-oriented development issue focusing on short-term economic prospects.

ACKNOWLEDGMENTS

The author would like to thank C. Michael Hall for the discussions of policy-making and sustainability in tourism and the reviewers and editor for their constructive comments. The study was originally funded by the Academy of Finland and it is part of the Academy's RELATE Centre of Excellence.

CONFLICTS OF INTEREST

The authors declare no conflict of interest.

REFERENCES

1. Britton, S. Tourism, capital and place: Towards a critical geography of tourism. *Environ. Plan. Soc. Place* **1991**, *9*, 451–478.
2. Butler, R. Tourism, environment, and sustainable development. *Environ. Conserv.* **1991**, *18*, 201–209.
3. Wolfe, R. Recreational travel: The new migration revisited. *Ont. Geogr.* **1982**, *19*, 103–122.
4. Meadows, D.H.; Meadows, D.; Randers, J.; Behrens, W.W., III. *The Limits to Growth:A Report for the Club Rome's Project on Predicament of Mankind*; Earth Island Limited: London, UK, 1972.
5. Britton, S. The political economy of tourism in the third world. *Ann. Tourism Res.* **1982**, *9*, 331–358.
6. Saarinen, J. Traditions of sustainability in tourism studies. *Ann. Tourism Res.* **2006**, *33*, 1121–1140.
7. World Commission on Environment and Development (WCED). *Our Common Future*; Oxford University Press: Oxford, UK, 1987.
8. Cater, E. Ecotourism in the third world: Problems for sustainable development. *Tourism Manag.* **1993**, *14*, 85–90.
9. World Tourism Organization (WTO). *Sustainable Tourism Development: Guide for Local Planners*; WTO: Madrid, Spain, 1993.
10. Blackstock, K.L.; White, V.; McCrum, G.; Scott, A.; Hunter, C. Measuring responsibility: An appraisal of a Scottish national park's sustainable tourism indicators. *J. Sustain. Tourism***2008**, *16*, 276–297.
11. Clarke, J. A framework of approaches to sustainable tourism. *J. Sustain. Tourism* **1997**, *5*, 224–233.
12. Miller, G. The development of indicators for sustainable tourism: Results of a Delphi survey of tourism researchers. *Tourism Manag.* **2001**, *22*, 351–362.
13. Valentin, A.; Spengenberg, J.H. A guide to community sustainability indicators. *Environ. Impact Assess. Rev.* **2000**, *20*, 381–392.
14. Mowforth, M.; Munt, I. *Tourism and Sustainability: A New Tourism in the Third World*; Routledge: London, UK, 1998.

15. Wheeller, B. Sustaining the ego. *J. Sustain. Tourism.* **1993**, *1*, 121–129.

16. Hughes, G. Tourism, Sustainability and Social Theory. In *A Companion to Tourism*; Lew, A., Hall, C.M., Williams, A., Eds.; Blackwell: Oxford, UK, 2004; pp. 498–509.

17. Poon, A. *Tourism, Technology and Competitive Strategies*; CAB International: Wallingford, UK, 1993.

18. Carrier, J. Protecting the environment the natural way: Ethical consumption and commodity fetishism. *Antipode* **2010**, *42*, 672–689.

19. Goodwin, H.; Francis, J. Ethical and responsible tourism: Consumer trends in the UK. *J. Vacat. Mark.* **2003**, *9*, 271–284.

20. Brickley, K., Black, R., Cottrell, S., Eds.; *Sustainable Tourism and Millennium Development Goals*; Jones & Bartlett Learning: Burlington, MA, USA, 2013.

21. Saarinen, J., Rogerson, C., Manwa, H., Eds.; *Tourism, Development and Millennium Development Goals*; Routledge: London, UK, 2013.

22. United Nations World Tourism Organization (UNWTO). *UNWTO's Declarationon Tourism and the Millennium Goals: Harnessing Tourism for the Millennium Development Goals*; UNWTO: Madrid, Spain, 2006.

23. World Bank. *Transformation through Tourism: Development Dynamics Past, Present and Future (Draft)*; World Bank: Washington, DC, USA, 2012.

24. Goodwin, H. Contemporary policy debates: Reflections on 10 years of pro-poor tourism. *J. Policy Res. Tourism Leis. Events* **2009**, *1*, 90–94.

25. Sharpley, R. Responsible Tourism: Whose Responsibility? In *The Routledge Handbook of Tourism and Environment*; Holden, A., Fennell, D., Eds.; Routledge: London, UK, 2013; pp. 382–391.

26. Rutherford, S. *Governing the Wild: Ecotours of Power*; University of Minnesota Press: Minneapolis, MN, USA, 2011.

27. Smith, D.M. Geography and ethics: A moral turn? *Prog. Hum. Geogr.* **1997**, *21*, 583–590.

28. Lawson, V. Geographies of care and responsibility. *Ann. Assoc. Am. Geogr.* **2007**, *97*, 1–11.

29. Lawson, V. Instead of radical geography, how about cariung geography? *Antipode* **2009**, *41*, 201–213.

30. Massey, D. Geographies of responsibility. *Geogr. Ann. B Hum. Geogr.* **2004**, *86*, 5–18.

31. Sin, H.L. Who are we responsible to? Local's tales of volunteer tourism. *Geoforum* **2010**, *41*, 983–992.

32. McCool, S.; Butler, R.; Buckley, R.; Weaver, D.; Wheeller, B. Is concept of sustainability utopian: Ideally perfect but impracticable? *Tourism Recreat. Res.* **2013**, *38*, 213–242.

33. Butler, R. Sustainable tourism: A state-of-the-art review. *Tourism Geogr.* **1999**, *1*, 7–25.

34. Hunter, C.J. On the need to re-conceptualize sustainable tourism development. *J. Sustain. Tourism* **1995**, *3*, 155–165.

35. Liu, Z. Sustainable tourism development: A critique. *J. Sustain. Tourism* **2003**, *11*, 459–475.

36. Sharpley, R. Tourism and sustainable development: Exploring the theoretical divide. *J. Sustain. Tourism* **2000**, *8*, 1–19.

37. Swarbrooke, J. *Sustainable Tourism Management*; CAB International: Wallingford, UK, 1999.

38. Sharpley, R. *Tourism Development and the Environment: Beyond Sustainability?* Earthscan: London, UK, 2009.

39. Lélé, S. Sustainable development: A critical review. *World Dev.* **1991**, *19*, 607–621.

40. Spangenberg, J.H. Will the information society be sustainable? Towards criteria and indicators for sustainable knowledge society. *Int. J. Innov. Sustain. Dev.* **2005**, *1*, 85–102.

41. Hunter, C.; Green, H. *Tourism and the Environment: A Sustainable Relationship?* Routledge: London, UK, 1995.

42. United Nations Environment Programme (UNEP). *Making Tourism More Sustainable: A Guide for Policy Makers*; UNEP: Paris, France, 2005.

43. Holden, A. *Environment and Tourism*; Routledge: London, UK, 2008.

44. Jones, S.B. Mining and tourists towns in the Canadian Rockies. *Econ. Geogr.* **1933**, *9*, 368–378.

45. McMurray, K.C. The use of land for recreation. *Ann. Assoc. Am. Geogr.* **1930**, *20*, 7–20.

46. Getz, D. Capacity to absorb tourism: Concepts and implications for strategic planning. *Ann. Tourism Res.* **1983**, *10*, 239–263.

47. Lucas, B. Wilderness perception and use: The example of the Boundary Waters Canoe Area.*Nat. Resour. J.* **1964**, *3*, 394–411.

48. Stankey, G.; McCool, S. Carrying capacity in recreational settings: Evolution, appraisal and application. *Leis. Sci.* **1984**, *6*, 453–473.

49. Wagar, A.J. The carrying capacity of wild lands for recreation. *For. Sci. Monogr.* **1964**, *7*, 1–24.

50. Wagar, A.J. Recreational carrying capacity reconsidered. *J. For.* **1974**, *72*, 274–278.

51. Butler, R. Carrying Capacity in Tourism. In *Tourism Research: A 20-20 Vision*; Pearce, D., Butler, R., Eds.; Goodfellow Publishers: Oxford, UK, 2010; pp. 53–64.

52. Butler, R.W. The concept of carrying capacity for tourism destinations: Dead or merely buried? *Progr. Tourism Hospit. Res.* **1996**, *2*, 283–293.

53. Tribe, J.; Font, X.; Grittis, N.; Vickery, R.; Yale, K. *Environmental Management for Rural Tourism and Recreation*; Cassell: London, UK, 2000.

54. Lindberg, K.; McCool, S.; Stankey, G. Rethinking carrying capacity. *Ann. Tourism Res.* **1997**,*24*, 461–465.

55. McCool, S.; Lime, D. Tourism carrying capacity: Tempting fantast or useful reality. *J. Sustain. Tourism* **2001**, *9*, 372–388.

56. Wall, G. Cycles and capacity: Incipient theory of conceptual contradiction? *Tourism Manag.***1982**, *3*, 188–192.

57. Mathieson, A.; Wall, G. *Tourism: Economic, Physical and Social Impacts*; Longman: New York, NY, USA, 1982.

58. Buckley, R. Ecological indicators of tourism impacts in parks. *J. Ecotourism* **2003**, *2*, 54–66.

59. Hammitt, W.E.; Cole, D.N. *Wildland Recreation: Ecology and Management*; Wiley and Sons: New York, NY, USA, 1997.

60. Taylor, J.P. Authenticity and sincerity in tourism. *Ann. Tourism Res.* **2001**, *28*, 7–26.

61. Collins, A. Tourism development and natural capital. *Ann. Tourism Res.* **1998**, *26*, 98–109.

62. McKercher, B. Unrecognised threat to tourism: Can tourism survive sustainability. *Tourism Manag.* **1993**, *14*, 131–136.

63. Weaver, D. Can sustainable tourism survive climate change? *J. Sustain. Tourism* **2011**, *19*, 5–15.

64. Burns, P. Paradoxes in planning: Tourism elitism or brutalism? *Ann. Tourism Res.* **1999**, *26*, 329–348.

65. Butler, R. The concepts of a tourist area cycle of evolution: Implications for management of resources. *Can. Geogr.* **1980**, *24*, 5–12.

66. Butler, R. Modelling Tourism Development: Evolution, Growth and Decline. In *Tourism, Development and Growth: The Challenge of Sustainability*; Wahab, S., Pigram, J., Eds.; Routledge: London, UK, 1997; pp. 109–125.

67. Martin, B.S.; Uysal, M. An examination of relationship between carrying capacity and the tourism life cycle: Management and policy implications. *J. Environ. Manag.* **1990**, *31*, 327–333.

68. Baum, T. Taking the exit route: Extending the tourism area life cycle model. *Curr. Issue. Tourism* **1998**, *1*, 167–175.

69. Murphy, P. Community driven tourism planning. *Tourism Manag.* **1988**, *9*, 96–104.

70. Timothy, D.J.; White, K. Community-based ecotourism development on the periphery of Belize. *Curr. Issues Tourism* **1999**, *2*, 226–242.

71. Jamal, T.; Stronza, A. Collaboration theory and tourism practice in protected areas: Stakeholders, structuring and sustainability. *J. Sustain. Tourism* **2009**, *17*, 169–189.

72. Selin, S. Developing a typology of sustainable tourism partnerships. *J. Sustain. Tourism* **1999**,*7*, 260–273.

73. Blaikie, P. Is small really beautiful? Community-based natural resource management in Malawi and Botswana. *World Dev.* **2006**, *34*, 1942–1957.

74. Scheyvens, R. Ecotourism and the empowerment of local communities. *Tourism Manag.* **1999**,*20*, 245–249.

75. Williams, P.W.; Gill, A. Addressing Carrying Capacity Issues in Tourism Destinations through Growth Management. In *Global*

Tourism; Theobald, W., Ed.; Elsevier-Science: Burlington, MA, USA, 2005; pp. 194–212.

76. Duffy, R. *A Trip Too Far: Ecotourism, Politics and Exploitation*; Earthscan: London, UK, 2002.

77. Hall, C.M. *Tourism and Politics: Policy, Power and Place*; John Wiley and Sons: Chichester, UK, 1994.

78. Stronza, A. The economic promise of ecotourism for conservation. *J. Ecotourism* **2007**, *6*, 170–190.

79. Bryant, R.L.; Wilson, G.A. Rethinking environmental management. *Prog. Hum. Geogr.* **1998**,*22*, 321–343.

80. Hughes, G. The cultural construction of sustainable tourism. *Tourism Manag.* **1995**, *16*, 49–59.

81. Redcliffe, M.; Woodgate, G. Sustainability and Social Construction. In *The International Handbook of Environmental Sociology*; Redcliffe, M., Woodgate, G., Eds.; Edward Elgar: Cheltenham, UK, 1997; pp. 55–67.

82. McCool, S. Constructing partnerships for protected area tourism planning in an era of change and messiness. *J. Sustain. Tourism* **2009**, *17*, 133–148.

83. Proctor, J.D. The social construction of nature: Relativist accusation, pragmatist and critical realist responses. *Ann. Assoc. Am. Geogr.* **1998**, *88*, 352–376.

84. Reed, M. Power relations and community-based tourism planning. *Ann. Tourism Res.* **1997**, *3*, 566–591.

85. Scheyvens, R. *Tourism for Development: Empowering Communities*; Prentice Hall: Harlow, UK, 2002.

86. Barr, S.; Devine, W.P. Resilient communities: Sustainabilities in transition. *Local Environ.***2012**, *17*, 535–532.

87. Wall, G. Sustainable Tourism—Unsustainable Development. In *Tourism, Development and Growth: The Challenge of Sustainability*; Wahab, S., Pigram, J., Eds.; Routledge: London, UK, 1997; pp. 33–49.

88. Hardy, A.; Beeton, R.J.S.; Pearson, L. Sustainable tourism: An overview of the concept and its position in relation to conceptualisations of tourism. *J. Sustain. Tourism* **2002**, *10*, 475–496.

89. Adger, W.N. Social and ecological resilience: Are they related? *Prog. Hum. Geogr.* **2000**, *24*, 347–364.

90. Pelling, M. *Adaptation to Climate Change: From Resilience to Transformation*; Routledge: London, UK, 2011.

91. Kajan, E.; Saarinen, J. Tourism, climate change and adaptation: A review. *Curr. Issues Tourism* **2013**, *16*, 167–195.

92. Inskeep, E. *Tourism Planning: An Integrated and Sustainable Development Approach*; Van Nostrand Reinhold: New York, NY, USA, 1991.

93. Gössling, S. Sustainable tourism development in developing countries: Some aspects of energy use. *J. Sustain. Tourism* **2000**, *8*, 410–425.

94. Gössling, S.; Hansson, C.B.; Hörstmeier, O.; Saggel, S. Ecological footprint analysis as a tool to assess tourism sustainability. *Ecol. Econ.* **2002**, *43*, 199–211.

95. Ringer, G. Introduction. In *Destinations: Cultural Landscapes of Tourism*; Ringer, G., Ed.; Routledge: London, UK, 1998; pp. 1–16.

96. Getz, D. Resort-centred tours and development of the rural hinterland: The case of Cairns and the Atherton Tablelands. *J. Tourism Stud.* **1999**, *10*, 23–34.

97. Eagles, P.F.J. Governance of recreation and tourism partnerships in parks and protected areas. *J. Sustain. Tourism* **2009**, *17*, 231–248.

98. Resilient People—Resilient Planet: Future Worth Choosing. Available online: http://www.un.org/gsp/sites/default/files/attachments/GSP_Report_web_final.pdf (accessed on 6 November 2013).

99. Scott, D.; Hall, C.M.; Gössling, S. *Climate Change and Tourism: Impacts, Adaptation and Mitigation*; Routledge: London, UK, 2012.

100. Tourist Operators Initiative to Sustainable Tourism. Available online: http://www.toinitiative.org/ (accessed on 6 November 2013).

101. Orams, M. The effectiveness of environmental education: Can we turn tourists into "greenies"? *Progr. Tourism Hospit. Res.* **1997**, *3*, 295–306.

102. Buckley, R. Sustainability reporting and certification in tourism. *Tourism Recreat. Res.* **2012**,*37*, 85–90.

103. Palacio, V.; McCool, S. Identifying ecotourists in Belize through benefit segmentation: A preliminary analysis. *J. Sustain. Tourism* **1997**, *5*, 234–243.

104. Swarbrooke, J.; Horner, S. *Consumer Behavior in Tourism*; Butterworth Heinemann: Oxford, UK, 1999.

105. Buckley, R. An ecological perspective on carrying capacity. *Ann. Tourism Res.* **1999**, *26*, 705–708.

106. Saarinen, J. The regional economics of tourism in Northern Finland: The socio-economic implications of recent tourism development and future possibilities for regional development. *Scand. J. Hospit. Tourism* **2003**, *3*, 91–113.

107. Kehittämisuunnitelman loppuraportti: Levi3. Available online: http://www.kideve.fi/sites/kideve.fi/files/Asiakirjat/levi3_raportti.pdf (accessed on 6 November 2013). (in Finnish).

108. Hall, C.M. *Tourism Planning*; Prentice-Hall: Harlow, UK, 2008.

109. Cohen, E. Authenticity, equity and sustainability in tourism. *J. Sustain. Tourism* **2002**, *10*, 267–276.

110. Hall, C.M.; Lew, A.A. The Geography of Sustainable Tourism: Lessons and Prospects. In*Sustainable Tourism: A Geographical Perspective*; Hall, C.M., Lew, A.A., Eds.; Longman: New York, NY, USA, 1998; pp. 199–203.

111. Redcliffe, M. *Sustainable Development: Exploring the Contradictions*; Methuen: London, UK, 1987.

112. Bianchi, R. Tourism restructuring and the politics of sustainability: A critical view from the European periphery (The Canary Islands). *J. Sustain. Tourism* **2004**, *12*, 495–529.

113. Dean, M. *Governmentality*; SAGE Publications Ltd.: London, UK, 1999.

114. Rhodes, R.A.W. The new governance: Governing without government. *Polit. Stud.* **1996**, *44*, 652–667.

115. Jessop, B. Government and Governance. In *Handbook of Local and Regional Development*; Pike, A., Rodriquez-Pose, A., Tomaney, J., Eds.; Routledge: London, UK, 2010; pp. 239–248.

116. Jamal, T. Resiliency and Uncertainty in Tourism. In *The Routledge Handbook of Tourism and Environment*; Holden, A., Fennell, D., Eds.; Routledge: London, UK, 2013; pp. 505–520.

117. McKercher, B.; Prideaux, B.; Cheunga, C.; Law, R. Achieving voluntary reductions in the carbon footprint of tourism and climate change. *J. Sustain. Tourism* **2010**, *18*, 297–317.

118. Fennell, D. *Tourism Ethics*; Channelview: Clevedon, UK, 2006.

119. Holden, A. In need of new environmental ethics for tourism. *Ann. Tourism Res.* **2003**, *30*, 94–108.

120. Macbeth, J. Towards an ethics platform for tourism. *Ann. Tourism Res.* **2005**, *32*, 962–984.

121. Scheyvens, R. *Tourism and Poverty*; Routledge: London, UK, 2009.

122. Hall, C.M. Pro-poor tourism: Do "tourism exchanges benefit primarily the countries of the South"? *Curr. Issues Tourism* **2007**, *10*, 111–118.

Index

S

T

U

W